RENEWALS 691-4574

DATE DUE

OCT 13			
DEC 9			
NOV 12			
NOV 12 2008			

Demco, Inc. 38-293

Traditions of Inquiry

TRADITIONS OF INQUIRY

JOHN BRERETON

New York   Oxford
OXFORD UNIVERSITY PRESS
1985

Oxford University Press

Oxford London New York Toronto
Delhi Bombay Calcutta Madras Karachi
Kuala Lumpur Singapore Hong Kong Tokyo
Nairobi Dar es Salaam Cape Town
Melbourne Auckland

and associated companies in
Beirut Berlin Ibadan Mexico City Nicosia

Library of Congress Cataloging in Publication Data
Brereton, John C.
   Traditions of inquiry.
   Bibliography: p.
   Contents: Introduction—Barrett Wendell /
Douglas—Fred Newton Scott / Donald Stewart—
[etc.]
   1. English language—Rhetoric—Study and teaching—
United States—History—20th century—Addresses, essays,
lectures. 2. Educators—United States—Addresses,
essays, lectures. I. Brereton, John C.
PE1405.U6T73 1985   808'.042'071173   84-27307
ISBN 0-19-503549-6

Printing (last digit): 9 8 7 6 5 4 3 2 1
Printed in the United States of America

ACKNOWLEDGMENTS

Some sections of Ann Berthoff's chapter on I. A. Richards first appeared in "I. A. Richards and the Philosophy of Rhetoric," *Rhetoric Society Quarterly* 10 (Fall 1980), and "I. A. Richards and the Audit of Meaning," *New Literary History* 14 (1982). Permission by the editors of each journal to reprint is gratefully acknowledged.

The editor particularly wishes to thank Ruth Ray of Wayne State University and Barbara Mirel of the University of Michigan for useful suggestions about the manuscript. Alice and Bud Trillin provided special help when it was most needed, as did John Wright, Carolyn McMahon, Joan Hartman, and Daniel Dixon.

The editorial staff at Oxford University Press has provided exemplary assistance to this project. My editor, Curtis Church, supplied just the right mixture of wise counsel and enthusiasm, while Henry Krawitz and Laura Brown provided expert guidance.

In acknowledgments it is usual for editors to thank their spouses for encouragement and support throughout the whole writing process. But I want to thank my wife, Virginia Lieson Brereton, for a good deal more. Since she is a writer, an experienced composition teacher, and a specialist in the history of American higher education, I was able to profit from her extraordinary editorial skills and her knowledge of the context in which composition teaching and research took place.

Heath, Massachusetts   J. B.
September 1984

CONTENTS

INTRODUCTION

Just two decades ago a book on the history of college composition would not have been written—or if some hardy soul had compiled such a volume, it would have appeared quietly and found a very small audience of antiquarians. Today all that has changed. The popular press has discovered what professors have known all along: Students entering college have great difficulty with all phases of writing, including coming up with something intelligent to say, gauging the needs of their readers, and meeting the minimum demands of edited English prose. We are in a writing crisis, and so composition instruction has taken on greater importance.

One result of this new importance is that composition is slowly becoming recognized as a distinct discipline, a subject closely allied with English, to be sure, but with its own aims, methods, and history. Some composition teachers and researchers are rediscovering composition's ties with rhetoric, both classical and modern; others are connecting writing research with work in the social sciences, including linguistics, sociology, and psychology. Still others look to literary theory for guidance, especially since literary theory has grown more receptive toward rhetoric and less dependent upon a canon of great authors. Composition research and pedagogy are rapidly developing, partly as a result of the external stimulus of a writing crisis and partly due to the perception that since the major share of an English department's work is in the teaching of writing, those hired to do it should have some specialized knowledge and professional training.

In the atmosphere of the 1980s, then, a book on composition's past can serve two distinct purposes: It can demonstrate how some important figures taught writing successfully in their own time, and it can provide current teachers and researchers with some of the historical background of their emerging discipline. Since the teaching of English composition has been, by far, the largest single enterprise in American higher education, it is time that people doing it understood something of their subject's past.

The eight figures discussed in this book influenced writing research and instruction in very different ways. The earliest, *Barrett Wendell* (1855–1921), one of Harvard's "great teachers" at the turn of the century, represents the high point (or, some would say, the degenerate end) of one strain of nineteenth-century writing instruction, with its emphasis on the surface features of style. *Fred Newton Scott* (1860–1931), the most prominent composition expert of the early twentieth century, headed the University of Michigan's famous rhetoric program, served as president of the Modern Language Association, helped found the National Council of Teachers of English, and wrote a series of widely used textbooks. *I. A. Richards* (1893–1979) and *Kenneth Burke* (b. 1897), both rhetoricians and literary theorists, provided a rich source of thinking about writing research and instruction. (And Richards, throughout his long career, also devoted himself to one of the most demanding and least glamorous fields of teaching and research, English as a Second Language.) *Sterling Andrus Leonard* (1888–1931) trained teachers at the University of Wisconsin, where he developed an innovative writing curriculum based on Deweyan principles and conducted some of the most important empirical research on language usage. *Theodore Baird* (b. 1901) of Amherst College published little but devised a brilliant writing program that helped shape many others around the country. *Richard Braddock* (1920–74) ran a large writing program at the University of Iowa; and through his publishing and editing activities for the National Council of Teachers of English he influenced a whole generation of composition researchers. *Mina Shaughnessy* (1927–79) worked with Open Admissions students at the City College of the City University of New York in the 1970s and spoke eloquently of the problems and potential of new learners in postsecondary education. Taken together, these eight individuals demonstrate some of the very diverse careers that have been possible for those interested in composition and rhetoric. They still exert their influence over the present generation of practitioners; they are people to confront, to agree with or argue against, to follow or to revolt against.[1]

Although these eight figures shared many concerns, it would be inaccurate to think of them as representing some sort of "great tradition." Interconnections do abound: Leonard was Scott's student and Richards' friend, and Baird's Amherst curriculum derives, at least in part, from Richards' writings. But Wendell does not "lead" to Scott, who "leads" to Leonard, who "leads" to Richards, who "leads" to Burke, and so on. Composition has been too protean a field—and the practitioners too varied—for them to encompass a single, unbroken line of research or teaching. Indeed, it is probable that many of the eight individuals would have been personally and intellectually incompatible.

Though the eight figures in this book demonstrate great diversity, they by no means represent every aspect of writing instruction in American colleges during the twentieth century.[2] For one thing, the individuals included here were exceptional in that they held senior positions at well-known colleges and universities. Most composition work has been carried out without fanfare, much of it by part-time instructors or teaching assistants. A disproportionate share has traditionally been done by women, usually at low salaries and all too often without the status of regular faculty appointments. For most of this century the economic reality at research university English departments has been that freshman composition brings in the student credit hours that make it possible to hire literature specialists. And these specialists teach graduate students, who in turn teach freshman composition courses. Thus, although composition has always been the engine driving the whole operation, most senior faculty members have been able to devote themselves to other pursuits. Some would say that, as a result, colleges have received the level of composition teaching they deserved, given the low regard displayed toward both teachers and students, and the almost complete divorce between a department's literary research and its teaching of writing. The eight figures included here, exceptional though they may be, demonstrate that absolutely first-rate work has often been possible despite difficult circumstances.

If there was one quality that characterized most of these eight individuals it was their conception of writing as a demanding intellectual process, one that encompassed everything from determining what to write about to selecting the best way to present it. They rejected the limited notion of rhetoric embodied in so much nineteenth-century teaching, the rote learning of accepted usage and elegant rhetorical devices. For them the making of meaning and the transaction between writer and reader were central to composition; and though correctness and style were important, they were clearly not primary. To people like Burke, Richards, Leonard, and Baird, composition was the working out of thought on paper; the interplay of language, ideas, and feeling; the shaping of meaning to create communication. It is this larger sense of writing as intimately bound up with reading and thinking that separates the best researchers and teachers from their less inspired colleagues.[3]

The accomplishments of the eight figures discussed in this book (and of a few other prominent figures who made significant contributions) should not blind us to the reality concerning composition in this century. For the most part, writing has been poorly taught and badly

researched. Demonstrations of the inadequacy of instruction and com-
plaints about the low level of research are a recurring theme in the
work of Leonard, Richards, Braddock, and Shaughnessy. Sometimes it
has seemed as if most writing teachers have had no real discipline of
their own, no body of knowledge to constitute the center of their study.
Composition specialists—even successful ones—have at times fostered
this notion by continually seeking something new and by greeting each
different approach—"scientific" research, New Criticism, structural linguistics, General Semantics, information theory, transformational
grammar, cognitive psychology, brain hemisphere studies—as the key
to the promised land. Thus, almost every generation has witnessed a
revolution in thinking about writing—a sure sign that the field has
lacked an overarching theory that would enable scholars to assimilate
the valuable aspects of new approaches as developments of, and additions to, an already well-defined discipline. All disciplines are subject to
fads, of course, but composition's fate has been worse than most: The
bankruptcy of the traditional error-hunting approach embodied in
handbooks has always been so evident that every promising new idea
was eagerly embraced. But these newer approaches have been just as
quickly discarded when something more attractive came along. The
name of the discipline's major professional organization bears ironic
testimony to this trend: The Conference on College Composition and
Communication (CCCC) owes the last part of its name to the "communication" approach popular in the late 1940s, when the CCCC was
founded, but which vanished soon afterward.

The strong belief in the panaceas promised by new approaches derives in part from the uncomfortably close connection between research
and classroom practice. Writing specialists have always been haunted by
the question of what to do Monday morning, so much so that their
research issues have often been determined by a pressing need to make
practical decisions. Although research shaped by classroom and administrative requirements has helped composition specialists avoid some of
the more recondite scholarly excesses of their colleagues in English
departments, the forced linkages among administration, research, and
teaching have rarely been conducive to slow, scholarly reflection or to a
thorough investigation of the writing process.

A perfect example of both the successes and the pitfalls of the practical emphasis in composition research is found in Mina Shaughnessy's
brilliant *Errors and Expectations,* which grew directly out of her realization that her Open Admissions students would fail out of college unless
they learned to write error-free prose in a great hurry. Faced with what
she termed students "at the eleventh hour of their educational lives,"
Shaughnessy wrote her book to help teachers produce immediate results, all the while acknowledging the obvious point that learning to

write is a slow, developmental process that takes place over a period of years. Yet so great was the desire for solutions that Shaughnessy's book has been used to help plan elementary and high school writing curriculums, entirely ignoring her explicit warnings that the book was about college freshmen in basic writing courses, not a curriculum guide for situations where a more thorough and consequently more effective approach to writing would be suitable.

Some of the same impetus toward practically oriented research lay behind other significant publications, notably I. A. Richards' celebrated *Practical Criticism,* which grew out of his discovery that otherwise well-educated British students could not make any sense of modern poetry. Sterling Leonard's *Current English Usage* resulted from his exasperation with composition textbooks that prescribed a complex, outmoded system of usage that effectively served as a barrier to student progress. Indeed, given the practical needs that have constantly faced writing specialists, it comes as no surprise that one of the most common forms of publication has been the textbook. Six of the authors represented here—Wendell, Scott, Richards, Leonard, Baird, and Braddock—either wrote or edited composition textbooks; Burke began a text but abandoned it; and Shaughnessy would probably have completed one (she went so far as to sign a contract) had she lived. For rhetoricians confronted with the need to translate knowledge into practice, textbooks have always been one of the best ways to embody their research in a form that could have a direct impact.

Though textbooks constituted a considerable portion of a composition specialists' publications, a large body of significant scholarship appeared in other forms as well: monographs, articles, edited collections of essays, research reports, curriculum guides, and teachers' manuals. Some of these have not always been recognizable as important contributions, given the humanistic bias favoring the scholarly monograph or learned article by an individual scholar. Educationalists too often display a tendency to view composition research narrowly. The journal Richard Braddock founded, *Research in the Teaching of English,* has from its inception advocated a rigorous social-scientific approach to research, as evidenced by its choice of the American Psychological Association (APA) format and its emphasis upon statistical reports replete with charts and graphs. Although such a journal came as a welcome antidote to the what-I-do-in-my-class testimonials published elsewhere in the 1950s and 1960s, its title and format should not suggest that it alone embodies the best or only acceptable type of research. Burke's theories of rhetoric, Leonard's investigations of proper usage, and the thinking embodied in Baird's Amherst writing curriculum all represent serious scholarship about composition; and though these individuals do not share a single approach, they have produced some of the most fruitful research the century has seen. Future contributions to the field of composition may well be made in many different forms—computer software or film come to mind—and will no doubt need to be recognized as such.

The history of composition in American colleges has yet to be written, but when it is it will draw heavily upon the work of the eight individuals discussed in this book. The inclusion of these eight profiles in one volume will, I trust, encourage others to be more curious about their discipline's past. I hope these biographies will prompt the reader to take a closer look at the activities of these eight figures, and perhaps even to emulate both the zest and commitment they brought to one of the most fascinating studies of all, the complex and endlessly intriguing process of determining what to say and how to say it well, which is what composition at its best will always be about.

Notes

1. Since it seemed important to look at an entire career, people still very active in the field were excluded; thus, six of the individuals are deceased, while two, Baird and Burke, are retired.

2. Others who had an extraordinary impact on the field include Porter Perrin, Charles Carpenter Fries, Alfred Marckwardt, Albert Kitzhaber, and Robert Pooley, to name just a few individuals from the recent past. I leave it to readers to determine which currently active researchers and teachers merit inclusion in such a list.

3. Wendell and Shaughnessy were partial exceptions to this approach, the former because his notion of composition was limited to surface style, and the latter because her research of necessity dealt with the causes of error among students in remedial English classes.

Traditions of Inquiry

BARRETT WENDELL

by Wallace Douglas

Barrett Wendell (1855–1921) taught various sorts of English at Harvard between 1880 and 1917. His appointment came about as the result of a chance meeting with Adams Sherman Hill, then the Boylston Professor of Rhetoric, probably in the early spring of 1880, while Wendell was in Boston for the funeral of a college classmate. Answering Hill's conventional questions, Wendell said that he was reading law, that he didn't like it, and that he would prefer any other sort of job, "Even yours." Years later (1918), in a memoir of his father that he wrote for his children, Wendell masked that last, graceless remark with "I am said to have answered." But, in fact, it is typical of his penchant for "astonishing utterances which seemed to delight him in proportion to their capacity to make his hearers 'sit up.' " And, of course, recording it, even though not quite acknowledging it, is typical of "the 'character' he was," or had made of himself (Howe, pp. 4, 36, 37). "Somehow," Wendell continued, "Somehow the incident stuck in [Hill's] memory." And in October 1880, he proposed to President Eliot that Wendell be hired to help him read themes in the sophomore division of "Prescribed Themes and Forensics."

Toward the end of his first academic year (1880–81), Wendell was told that the college did not have enough money to keep two instructors in English. Perhaps with some feeling of noblesse oblige, Wendell at once resigned, because the salary meant more to his fellow than to him. He took the next year off and tried to write a novel. Then early in October 1882, Eliot offered him "an annual appointment at twelve hundred." As the appointment was renewable ("annual") and involved "rather more work" than he had had, Eliot must have been expressing some sort of satisfaction with his performance in 1880–81, perhaps even some confidence in his future. Wendell seems to have been somewhat reluctant about going "back to Cambridge," but he told his wife that he would write Eliot at once, "undoubtedly accepting" (Howe, pp. 29, 37, 52).

He did accept and, what is more, survived. His promotions came decently, if not rapidly: assistant professor in 1888, professor in 1898. Two unsuccessful novels (in 1885 and 1887) could not have much advanced his case. But the Lowell Lectures in 1890 (published as *English Composition* in 1891) would have brought him to the attention of a larger public than the students to whom he had been purveying their substance for a decade. Holmes, Cabot Lodge, Shaler, Fiske, and C. W. Eliot himself had all preceded Wendell as Lowell lecturers. Amid such men and their lofty topics and with the credit conferred on him as the choice of Augustus Lowell, how could it have been supposed that in *English Composition* Wendell was but laying before his audience, "as simply and as broadly as [he] could, the theory of style to which ten years of study" and ten years of teaching Harvard undergraduates had led him? Yet, indeed, that is what Wendell in all his bravado, with his usual bravura, told his audience. Perhaps it was his biography of Cotton Mather (1891), which has been loyally praised (in 1926) by a former student and (in 1963) by the editor of a modern reprint, that gave Wendell his pass into the respectability of a professorship (Weeks, p. 96; Self, p. 45).

I could have written "safety" just as well as "respectability." For Wendell seems to have been somewhat neurasthenic, or, using his own words, to have had a nervous system that was "always a bit erratic," a constitutional weakness that he traced through [his] father and his [father's] mother to hers"; that is, to his great-grandmother, a Sherburne of Portsmouth, "who had a startling experience of interrupted personality about 1800" (Howe, p. 11). For the etiology of his own case, Wendell used "my probably hysterical paralysis," presumably a more up-to-date clinical term than "interrupted personality," or, more benignly, "nervous over-sensitiveness." Like others in the period, Wendell thought that "a great deal of modern illness is in some degree hysterical," so we should not exaggerate his harsh self-diagnosis. But the fact is that Wendell was subject to depressive episodes that were severe enough to require, if not medical intervention, at least escape from the anxiety-producing situations (Howe, pp. 17, 20, 72).

One of these, occurring when he was a schoolboy, made his parents decide that he "should no longer be exposed to the vexatious discipline of a large school." They put him into a small class being prepared for Harvard by a John Adams, a cousin of his mother (Howe, p. 17). Another episode, which he described as a "recurrence" of the paralysis, came upon him in 1873, in the spring term of his first college year. Rest and fresh air were prescribed, so the summer was spent in Swampscott, in the house of cousins of his Grandmother Barrett, the Addison Childs, "who then owned the whole neighborhood." The rest extended into the following year, beginning with a voyage as far as Rio de Janeiro, on a ship in which a friend, who "had also been out of condition," was acting as purser. The ship was sailing without passengers, so presumably Wendell, too, was taken along as a hand. In any case, by the time the ship had put into Rio for repairs, Wendell had regained enough energy and initiative to be able to decide to leave the ship and go off, alone, to his "beloved Europe" (Howe, p. 20).

For a young man of eighteen, lately beset by listlessness or, at least, inability to study, the subsequent travels are remarkable. (Indeed, they might be thought so even for one of today's footloose students.) Wendell began in Spain, did bits of Mediterranean Africa, spent weeks in Rome, "most of May" in Paris, went north up the Gulf of Bothnia as far as the Arctic Circle, "drove across" Scandinavia, and saw St. Petersburg, Moscow, Warsaw, and Berlin. He ended up in England, whence, "by the last side-wheel steamship—the *Scotia*—which ever crossed the Atlantic," he returned to Cambridge. Apparently, all the to-ing and fro-ing had strengthened his nerves enough so that he was able to finish Harvard, in President Lowell's gentle phrase, "standing high in his studies, though not among those at the top of his class, for his interest was rather literary than learned, and he had no ambition for rank as such" (Howe, pp. 20–21, 23).

For three years after graduation Wendell studied law in his fashion, first at Harvard, then, after not troubling to take finals, in the office of a family friend in New York, finally in Boston in the office of Shattuck, Holmes, and Munroe, where he found his distant cousin, Oliver Wendell Holmes, Jr, to be "one of the most charming men [he] had ever met" (Howe, pp. 29, 35, 36).

But apparently reading law suited him neither physically nor psychically. At Harvard he found the chairs hard, and his back muscles, which he thought had "always been rather weak," were constantly strained from reaching for, lifting, and carrying "heavy volumes" of legal lore, as Langdell's newfangled methods were beginning to require of students. Wendell was kept "constantly tired" by the physical effort (Howe, p. 36).

As Wendell remembered them, his father's letters at the time of the Harvard failure did not contain "even a syllable of reproach." The father accepted Wendell's rationalization that law school involved too much "abstract study" and that "studying in an office" might prove "less repugnant." In the event, Wendell was no better able to stand title searches and indexing court reports than boning up cases. The work, though "admirable" as discipline toward accuracy, "was solitary and far from stimulating to the imagination." By late winter of 1879, Wendell was in another "nervous depression," unable to write a piece he had promised to the *Lampoon,* feeling his self-control quite gone, sometimes indignant at his "weakness," at other times feeling completely abandoned to it and growing "weaker and weaker." He felt harsh self-contempt as he watched himself, "a sane, healthy-looking man of three-and-twenty who ought to be a gentleman, and as a gentleman have something approaching a respectable amount of character and determination, fading out into a miserable limp paper doll, which has not even the recommendation of being dressed in the fashion" (Howe, pp. 31, 36).

The figure of the paper doll suggests that, depression or no, Wendell was still able to turn a phrase for a friend. And perhaps in some part of himself he was enjoying a romantic agony. But the depression cannot have been wholly theatrical; or the doll, wholly decorative. For Wendell's father, who seems to have watched over his son's moods and humors with some care, noticing even just "the shadow of great trouble," now let him join a college classmate for a vacation in Florida. In Jacksonville he composed for a confidant a rather formal discussion of "the dawning sense of uselessness which saddens the lives of men" like them, who find themselves "with leisure and nothing in particular to think about." He added, "Life for its own sake is not to us worth living."

Can he have been wishing for a cause like those that had given meaning to the lives of the young men who had fought the Civil War and its preliminaries? Wendell had the political and social convictions of his time and class, but it is hard to see him as an activist. So perhaps "life for its own sake" only reflects the pickle he found himself in, abhorring business, yet unable to develop any toleration for, much less satisfaction from, the law. The rest of Wendell's composition is about easeful death and the considerations that prevent him from seeking it at once, namely, "the feelings of people whose lives I do not help make happy," and the "psychological post-mortem" that a suicide brings on (Howe, pp. 29, 32–33).

In the spring of 1879, Wendell became engaged to Edith Greenough, of Quincy, whose father was for many years president of the Trustees of the Boston Public Library. Howe says that the engagement caused the vanishing of two of the three causes of Wendell's depression: "imperfect health, uncongenial prospects, and a general lack of incentive to satisfactorily directed effort" (Howe, p. 34). He does not say which two. Presumably the link to the Greenoughs improved his "prospects," though as they must still have involved the law, it is hard to see them becoming more congenial. Perhaps the anticipation of marriage gave more satisfactory direction to his efforts, whatever they were.

In fact, though it may have lifted his depression, the engagement seems to have had little effect on his ability to get himself in hand. Apparently, to his surprise ("chagrin" was his word), he failed the bar examinations, the first candidate from the Shattuck firm ever to do so. According to Wendell, neither his father nor Shattuck expressed anything more than "regret" at his failure though he expected the former to be "furious" and the latter "resentful." Nor did anyone think to halt the preparations for the wedding, which occurred in Quincy on 1 June 1880. His father gave him a letter of credit for five hundred pounds, and the newlyweds had three months of travel in Europe. They returned to his parents in New York, and it was there that Wendell received Eliot's telegram asking him to come to Cambridge to discuss a job. "The telegram," he said, "decided my career; it also gratified my father as indicating, for the first time, that somebody thought me conceivably useful" (Howe, pp. 37, 38). Considering Wendell's connections, I find it hard to believe that his meeting with Adams Sherman Hill simply "stuck in [Hill's] memory" until, months later, he needed help with sophomore themes or that the appointment came "fortuitously," as Wendell said (Howe, pp. 37, 38), unless, of course, he was using "fortuitously" in the popular sense.

II

For Wendell, whose "personal stars in their courses fought against the law and practical affairs [and] on the side of letters," an academic berth was an obvious way out, especially one at Harvard. Indeed, Wendell proved to have all the skills necessary for high success within the Yard. His "oddities of temper and of manner," he once said, "chanced to interest [his] pupils" (Howe, p. 39). Very clearly, Wendell was not just a teacher, one destined to rub along to retirement, legendless, nicknameless, unmemorialized even in late night, boozy reminiscences at conclaves of scholars. As things turned out, Hill's rescuing appointment gave Harvard yet another of its Great Teachers. And what is it that sets Great Teachers apart from their humdrum fellows? The answer is *eccentricity*. To be a great—or at least a remembered—teacher, an individual must be capricious in garb, gesture, or speech, and—in Wendell's case anyway—outrageous in behavior.

Once when he was a visiting professor at Berkeley, in the summer of 1901, Wendell asked a class in composition "to state in writing what they expected to get out of the course," for Wendell a rather teacherly topic. According to the Fresno *Republican* (4 July 1901), "One feminine student wrote that she had 'come to the university lured by the fame of the great professor from Harvard,' whom she had long worshipped from afar off, 'to sit at his feet and gather inspiration from his gifted lips.' This extraordinary effusion Professor Wendell characterized as 'disgusting slop,' and said that he had never known a woman to make such a fool of herself on one page." The *Republican* thought Wendell impolite, but did not question the judgment itself (Howe, p. 126).

Of course, anyone brought up in the Harvard tradition of simple writing and honest expression (which goes all the way back to Edward Tyrell Channing, the Boylston Professor of Rhetoric between 1819 and 1851) would have found the student's words and sentiment rather overblown. In Wendell's case, the attitude would have been exacerbated by his general disapproval of women students. He expressed this in an 1898 report on relations between Radcliffe and Harvard. Women, he bluffly asserted, cannot provide the kind of "mental resistance" that men teachers can find in male students when they are more or less of an age with them. "In brief," he wrote, "a man who likes to teach women is in real danger of infatuation," meaning, of course, "in danger of becoming fatuous," not of having improper relations with women students. Twenty years later Wendell could still think of the report as his "most far-reaching work at Harvard" (Howe, pp. 87, 89). Needless to say, as man and professor, Wendell could not for a moment have entertained the possibility that the Berkeley writer might have intentionally pushed her language and sentiments over the edge into irony.

For nearly forty years Wendell strode—or tripped—about the Yard, wearing a cane and a beard, "twirling [his] watch-chain while addressing an audience," using "a staccato and much inflected speech that was eminently his own—not the utterance of Oxford, yet much more English in its effect than American." Late in life Wendell confessed that he had been little pleased when he had heard a recording of his voice (Howe, pp. 39–40, 274).

One of his students, writing a few months after his death (8 February 1921), tried to describe Wendell's style and effect as a teacher. Men of "flabby intellect," the man said, got little from Wendell.

> Prigs were shocked. A very few never recovered from the first surprise at his high-pitched voice and explosive speech as he strode heavily up and down the platform, twirling his watch-chain. They chose to consider him affected, thereby missing at the start one of the profoundest realities about the man—his contempt for affectation.

The writer goes on to say that Wendell taught the best students—those he might have called "the better sort," "men of one's kind," "the students we personally know, men of our own class"—"how to use their brains" (Howe, pp. 6, 75, 85, 86).

Even in the minute presented to the Harvard faculty after Wendell's death, there is mention of "eccentricities of voice and manner," of "humorous epigrammatic exaggerations," of "willingness to shock the straightlaced with what they could not but regard as levity and irreverence," of being easily imitated for mockery, of shooting off snap remarks about matters of public concern, which could be torn from their contexts to make good stories for Boston papers. Such conduct led many to "the impression that he was a light-minded aristocrat whose conduct was quite out of keeping with his official position" (Howe, pp. 83–84). The chief writer of the minute was Dean Briggs, who probably knew that Wendell would have been flattered and pleased to hear himself described as a "light-minded aristocrat"; then he would have known that he had played his part successfully.

There must have been something of a scamp in Wendell, Scaramouche without occasion for cowardice. He himself said "that a great part of whatever success [he had] as a teacher [was] the result of indiscretion"; and in the report of his death to the Colonial Society of Massachusetts, he was credited with "[a] faculty of committing judicious indiscretions."

Again according to Dean Briggs,

> His pupils knew that he kept nothing back, that he was never warily on his guard, that they had whatever was in [his] mind, and that the mind was fertilely original and bold. . . . It is doubtful whether any teacher or writer in America has equalled him in the quick and clear perception of literary relations, in the power of generalizing.

And, the dean concluded,

> Along with his deep-thinking seriousness men discovered his open-hearted simplicity, his genuine humility, and his singular loveableness. He died a man who had given new and just distinction to the University and to literary scholarship in America [Howe, p. 84 and note].

Perhaps the best summations of this quirky man are those by his college classmate Abbot Lawrence Lowell, who was president of Harvard when he wrote, and by his student and biographer, M. A. DeWolfe Howe. Wendell, Lowell said, was two men. "There was the real man, and [the man] he thought himself to be; and the former was the larger of the two." And Howe, accepting Lowell's judgment or diagnosis, warned readers of the biography to be "constantly aware that what the subject of this book really was, and what he thought himself and wished others to think him, were frequently at variance." Then expanding on Lowell, he noted that Wendell's life "covered the time when dual, and even multiple personalities were becoming objects of special scrutiny," as if to say that Wendell's "oddities of temper and of manner" were caused by the study of dissociation or even by the strange tales of Dr Jekyll (1886) or of Dorian Gray (1891). Furthermore, the point is made in a subordinate clause ("Throughout his life, which covered . . ."), as if Howe wanted to slip it in so that his readers would not need to notice its implications, at least not fully.

In any case, whatever may have lain behind the point, it fits well enough with one of Howe's conclusions about Wendell: "Throughout his life he was appearing in the guises—sometimes the disguises—which made him the 'character' he was . . ." The phrasing again suggests that Howe wanted to get in a psychiatric warrant for his interpretation, but at the same time to keep his readers from finding anything pathological in Wendell's sometimes violent swings of mood. I think he may have been unduly apprehensive. Wendell may well have been a cycloid personality. But if we put aside his "nervous prostrations," Wendell's moods were probably well within the normal range; at least he was not known for the kind of tantrums Kittredge was pleased to put on. (See Howe, pp. 3, 4, 39.) Wendell seems to have had another episode of "prostration" in 1915, which continued until his retirement in 1917 (Howe, pp. 268, 277, 278).

III

Early in *English Composition* (1891), the book he got out of his Lowell Lectures (1890), Wendell remarked that he knew no term "more precise than Style" to express "the whole subject he would be discussing (Wendell, p. 3). Then he defined style as "simply the expression of thought or emotion in written words . . ." And then, perhaps having noticed the hopelessly wide extension of his definition, he added that style "applies equally to an epic, a sermon, a love-letter, an invitation to an evening party" (p. 4). A bit later he used "a piece of style" to name the sort of thing he had under consideration, again helpfully specifying the abstraction, this time with "a poem, a book, an essay, a letter" (p. 7).

Obviously, in those passages both the term and its rather sketchy denotation float on the morass of literary theory and speculation that had proceeded from school people of the period, who were constantly trying to define the literariness of, say, Bacon's essays or Carlyle's histories. Wendell's references to the controversy legitimize, lend weight and gravity, without otherwise much affecting his discussion of composition teaching. Eventually, they prepared the way for his defense of composition teaching.

At the same time, of course, Wendell used his little lists of class members to simplify and domesticate a very complex problem. The descent from epic style, which his readers would be familiar with, to a relatively petrified social form ("an invitation to an evening party"), which his readers could accept as having style, would bring them comfortably to rest in their common experience. They could feel a frisson of knowing amusement at the unexpected linking of epic and invitation, which would suggest to them scholarship, severe analysis, novel thought, perhaps even a glancing blow at the fetish of the classics, all presented with the most fetching simplicity.

There is yet a third example of Wendell's pawky lecturing style within the opening movement of *English Composition*. Wendell presented it as an attempt to clarify the definition of style as "the expression of thought or emotion in written words." The definition, Wendell said, "brings us face to face with a trait which the art we are considering [that is, the English composition that is practiced in schools!] shares with the other arts of expression . . ." The addition of the class "arts of expression" is a little disorienting, and it is not immediately apparent whether Wendell used the words as a possibly elegant variation of "expression . . . in written words" or whether he intended to introduce a different and more sophisticated class. I suspect the former because, in specifying "arts of expression," Wendell yet again produced a list of class members that is conspicuous for heterogeneity, even beyond the other two. The class now includes "painting, sculpture, architecture, music," and, astonishingly, "those humbler arts, not commonly recognized as fine," such as bring into existence things like "a machine, a flower-pot, a sauce" (p. 4).

Surely, among Wendell's readers, if not his audience, there must have been some to wonder what he hoped to accomplish by reviving the oldest uses of *poiein* as a word for making in general even if most of them would have been wondering what common attributes were shared by machine-making, pot-making, and sauce-making. But perhaps Wendell's homely references would have teased his audiences out of nagging thought, leaving them entranced by his witty philosophizing. Of course, Wendell had an answer. The identifying property that unites flowerpots and epics, sauces and sonnets, machines and essays is that in all of them "the workman conceives something not yet in existence . . . and proceeds, by collaboration of brain and hand, to give it material existence" (p. 4). As it stands, the clause is syntactically related only to "the humbler arts." But it can be related to the fine arts by Wendell's opening statement that composition shares "an obvious trait" with the other arts of expression. Only "making," giving "material existence," can be construed as a trait shared by all the arts of expression. There remains, of course, the problem of differentiating sauce and sonnet making. Prudently Wendell never raises it, being content in his muddle of Greek thinking about art. Still, such classical resonances were all but essential to public discourse at the time.

Because Wendell put much energy into styling the character he believed himself to be, perhaps I may comment on his otiose "by collaboration of brain and hand." Most composition teachers I have known would have struck through the words on the grounds that their meaning is clear by implication of the main statement. How else could form be imposed on material? But, of course, deleting the words destroys the cadence of Wendell's sentence. Unmodified, "proceeds" is rather commonplace, a word too close to ordinary matters. And as a whole, "proceeds to give it material existence" is a little curt, comparatively accentless, especially with the run of syllables in "to give it." With "by collaboration of brain and hand," Wendell suspended the slight hurry of the predication, gave it a more definite rhythm, and not incidentally clothed the statement with whatever authority his audiences may have been accustomed to find in scientific terms.

Immediately after that taxonomic exercise, without any paragraph break or even a transitional word, Wendell inserted a sort of arietta on thought and emotion, of which style, it will be remembered, is the expression. At this point, they become "the substance of what style expresses." It is not clear how those words are related to "the expression of thought or emotion in written words." Perhaps Wendell inserted "substance" as an afterthought lest his audiences think him talking about no more than word order. (Anyone who reads much of Wendell is likely to fall into constructions like that "think him talking.")

In the remainder of the sentence, Wendell asserted that thought and emotion

> are things so common, so incessant in earthly experience, that we trouble ourselves to consider them as little as we bother our heads about the marvels of sunrise, of the growth of flowers or men, of the mystery of sin or death, when they do not happen to touch our pockets or our affections [p. 4].

Modern readers, if they manage to tolerate the overelaborated rhythm of the passage and the commonplace poetic clichés, may still wonder how to read the last clause. Oh, they'll recognize the trick in the sudden descent to what, in keeping with the style of the passage, I must call the earth earthy. And, briefly, they might be amused by it, though wondering if the phrase is not so full and soft as rather to muffle the intended effect. But after responding to Wendell's medium jinks, I think they would end up judging that Wendell had lost his idea under all the stucco of words. In plain English, he was saying no more than this: "The thoughts and emotions of human beings are so much a part of consciousness that they are never noticed except when they have money or themselves as object." Said thus, without the Wendellian flourishes, the statement is either truistic or palpably false—or incoherent, for "emotions" are hard to distinguish from "affections." The only effect of all the fine words is to mask, or at least half-mask, the emptiness of the thought. Wendell's way of performing led him straight and quickly into fine writing and trivial thinking.

IV

In *The Middle Span,* the second volume of *Persons and Places* (1945), Santayana says that Wendell "had no real distinction in himself; his mind and his attachments, like his speech, were explosive and confused; there was emotion, often deep emotion, but it broke out in ill-governed and uncouth ways." He longed for what he thought "the old colonial proprieties" and cultivated a passionate regard for "our tradition from eldest New England time." "In a word, [he] was a sentimentalist."

> Had he been thoroughly educated and a good Latinist like Dr. Johnson, he might have expressed and propagated his ideals to better purpose; as it was, his force spent itself in foam. He was a good critic of undergraduate essays, but not a fair historian or a learned man; and his books were not worth writing. He was useful in the College as a pedagogue, and there was moral stimulus in his original personality [p. 171].

The judgment is harsh, especially in its total dismissal of Wendell's books. At least the one on Shakespeare (1894) is worth attention, if only for the shrewd case Wendell makes for Goneril and Regan (pp. 296–299). I can find less to say for *A Literary History of America,* which is written in a style much too high for the book's own good and which also suffers from Wendell's habit of dwelling "on the birth or family of literary men," which Eliot called "a subject for regret, because it shows that he has not observed how quickly American men and women acquire not only the manners and customs but the modes of thought and speech, and the sentiments which prevail among 'ladies and gentlemen. . . . Wendell's frequent discourse on the subject of birth and descent seems snobbish in an American . . ." (James, vol. 2, pp. 134–135).

Santayana extended his harshness even to Wendell's famous "dailies," which he called "a voluntary exercise in writing, feeling, and judging of all things like a gentleman," and said that from them students "learned nothing except what to think about what [they] happened to know" (p. 172). Except at Yale, dailies have long existed only in the folk memory of the profession, but Yale's course catalogs listed the course Daily Themes between 1907 and 1977, and the course was revived in the winter of 1983 (Engle, p. 16). According to Santayana, Wendell's dailies were written on half-sheets of notepaper. They were turned in at Wendell's office on the morning of the day following composition. Overdue ones were never considered. Santayana says that Wendell read them in their hundreds (p. 172), but I suspect he exaggerates when he says that they were revised. So far as I know, the dailies were merely exercises in fluency and specificity and also a means of encouraging originality of perception. Wendell insisted that they not have titles; presumably, he wanted to keep up the fiction that they were just passing notes by men of some intelligence trying to find significance in the moon sighted from a window in Holworthy or in other "incidents and situations from common life."

It would be wrong to consider Wendell a creative power in the history of the teaching of composition when the teaching device for which he is chiefly remembered, if at all, is one that goes so counter to all that is supposed to have been learned about the "writing process" in the last few years. But, of course, there is *English Composition,* in which he set out his philosophy of composition or, more properly, style. *English Composition* (1891; rev. 1894. I am using a printing of 1899) is both a textbook and, at points, a rather impassioned apologia for a life spent in "the dreary work" of plodding through "ream upon ream of manuscripts that college students write in an effort to learn how to make themselves writers" (pp. 303, 304). It is Wendell's attempt to justify his feeling

> that the work so many of us are trying to do at Harvard College . . . the work of any earnest teacher of this subject—composition—that seems to most men so dull, is a work that may rightly claim a place in any system of education, no matter how high it hold its head [p. 306].

In the minute on Wendell's life and services, *English Composition* is described as "probably the most suggestive book ever written on composition." In 1921, just before Wendell's death (8 February), a man in Arkansas sent him the copy of *English Composition* that he had used at the University of Arkansas under one of Wendell's students. The man asked Wendell to autograph the book, adding

> I would not take $50 for this book, for it has been my constant companion for fifteen years. . . . When I went away to war I only took three books— Wendell's *English Composition,* Stevenson's *Essays,* and a New Testament [Howe, p. 79].

It had taken him, Wendell said, "the better part of ten years to think out, from a snarl of books and practical experiments, the very obvious principles," which, laid out in *English Composition* "as simply and broadly" as possible, constituted a "theory of style" (pp. 135, 308). Ten years of teaching had led him to find that "none" of the available textbooks "seemed quite simple enough for popular reading." That gave him "excuse for offering a new treatment of the subject." As, in the same place, he acknowledged his "obligation [i.e. debt] to the textbooks" of Hill, Bain, Genung, and "the late Professor McElroy" (prefatory "note"), presumably it was their work he proposed to simplify for the use of college students and self-improving laypersons. His acknowledgments suggest that, whatever the newness of his "treatment," Wendell's substance was well within the tradition, or he wanted it to appear so.

Wendell seems to have been unable to rid himself of the notion that a composition teacher is "a technical critic of style" who is concerned "largely with the correction of erratic [sic] detail." But then, even style is a better object of attention than the kind of trifling detail that, according to Wendell, absorbed his colleagues. In the ten years at Harvard that produced *English Composition,* Wendell's friends and colleagues seem to have treated him like a walking handbook. They asked him questions about the admissibility of "this word or that," about why he had once used "the apparently commercial phrase 'at any rate,'" about whether they should try for a high Anglo-Saxon count in their diction or go Latinate, about whether they should keep their sentences short or make them long (p. 1).

Not many people—Adams Sherman Hill and LeBaron Russell Briggs may have been exceptions—like to have themselves seen as proofreaders, connoisseurs of barbarisms, improprieties, and solecisms. Wendell tried to dispel that impression of himself by saying that, in the course of his preparatory ten years, he had come to see "more and more clearly that . . . the really important thing for one who would grasp the subject to master [it] is not a matter of detail at all, but a very simple body of general principles under which details readily group themselves" (p. 2). At the time "general principles" of composition were much discussed, though not discovered with any great success. And, in fact, *English Composition* contains a fair amount of material on "erratic detail," though of course it is not set in the form of rules; see the "Note for Teachers" that Wendell added in 1894 and also Chapter 2 of the text itself.

Like many composition teachers, Wendell had an outlining mind, which is not quite the same as an analytical mind. And his discussion presents him as finding his "very simple body of general principles," his "very few simple, elastic general principles" (p. 2) by an outline partitioning of the topic Style, or "the whole subject" of *English Composition* (p. 3). Except in his odd remark that thought and emotion are "the substance of what style expresses" (p. 4), Wendell seems to have been relatively uninterested in what is called "content." Like other textbook makers of the time, he stuck to the abstract vehicles of expression: words, sentences, paragraphs, and whole compositions. These he called "elements of style," and I suppose that, in a loose sort of way, the shape of a writer's paragraphs and sentences will help to constitute the effect of the writing and so may be called an "element." I am not at all clear about how a whole composition, composed of words, sentences, and paragraphs, can at the same time be an element of style along with its components.

Wendell said that a piece of style cannot exist without affecting or, to use the technical term, impressing the people who are in transaction with it. These impressions make it possible to say that a piece of style differs from another in more or less describable ways or has distinguishing, nameable qualities. Given that "each man's thought and emotion differs [sic] from every other man's," it must follow that the impressions will be as various as the persons who experience them. But for a simplified, popular book that is to be teachable, multifarious experience must be categorized, in this case according to what Wendell called the Qualities of Style. Three being a good number in pedagogy, he offered the proposition "that the undefined impression which any piece of style makes may always be resolved into three parts." He suggested that, with "a little consideration," his audiences would agree. Then he gave the three possibilities, beginning with a phrasing in common language. A person understands a piece of style or doesn't or is left uncertain. A person is interested, bored, or indifferent. A person is pleased, displeased, or in doubt as to which.

Appearing to explain this not exactly complicated idea, Wendell first moved into a more technical phrasing: "In short, every piece of style may be said to impress readers in three ways,—intellectually, emotionally, aesthetically; to appeal to their understanding, their feelings, their taste." Next he repeated his topic sentence: "Every quality that I know of may be reduced to one of three classes," which alone "are different enough to deserve distinct and careful consideration." Then he repeated the three qualities of style in adjectival form—"intellectual, emotional, and aesthetic"—presumably to get them into a form usable in descriptions or classifications of particular pieces of style. Finally, borrowing from Hill, Wendell named the qualities: Clearness for the intellectual, Force for the emotional, and Elegance for the aesthetic (pp. 7–8, also 193, 308–309). The thought does not much advance.

It is hard to believe that teachers once accepted talk about clearness "as the distinguishing quality of a style that cannot be misunderstood" (p. 194), about force "as the distinguishing quality of a style that holds the attention" (p. 236), and about elegance as "the distinguishing quality of a st/le that pleases the taste" (p. 272). But Self says that *English Composition* remained in use as a text until 1942 and suggests that it influenced many textbook writers of the era (p. 134). So perhaps it will be useful to follow Wendell's discussion of one of his famous terms—"clearness"—to see something of what teachers have been willing to accept as the organizing ideas of their mystery.

Wendell began by noticing that he did not understand the formulas of physics or some of the jargon of football, but that physicists and "sportsmen" can be equally "bewildered" by rhetorical terms. This fact led him to disclose to his audiences "that clearness is not a positive quality, but relative; that what may be perfectly clear to one man may be hopelessly obscure to another." For Wendell there followed the question of what kind of person, what fictive audience writers should imagine themselves addressing if they want to achieve a style that has clearness. The question, Wendell said, "admits of a pretty definite answer. A generally clear style is a style adapted to the understanding of the average man" (p. 194–196). In spite of the adjective, Wendell's average man seems to share not a little with Johnson's common reader. For he is "the permanent type of those simplest and broadest traits, of thought and emotion alike, which make the brotherhood of the human race" (p. 198).

It is something, I suppose, that Wendell went beyond the high classicism of his average man in all his grand humanity to state with some firmness that a writer's first need is "to realize the range and limit of his reader's information" and then to put what he called "the precise question" of "what the average man can be expected to find vague or ambiguous or obscure" (pp. 201, 209); the adjectives refer to the three kind of unclearness that Wendell had developed in his earlier discussion (pp. 202 ff.). At least the two points suggest that the control of a writer's planning should lie with some empirical information, not with what can be come to by putting the "headings" of a topic on bits of paper or cards, then shuffling them around to find an order and an argument (pp. 165, 211). But, of course, the average man is as much a token in a circular argument as ever was Johnson's common reader, and only in a classroom can *a* writer have *a* reader, the "range and limit" of whose information can be determined. Besides, note the problems that must follow when children and young people are categorized as writers, and distinctions are collapsed between their needs as learners—and of more than just writing—and the needs and habits of writers.

At this point, a modern reader may feel compelled to ask, perhaps in some exasperation, whether Wendell ever got beyond the surface characteristics ("qualities" in his lexicon) of pieces of style. As a matter of fact, he did, though never for long, and never with any sense that the substance of thought and emotion—what style expresses—is as interesting as the elements and qualities of the vehicles of expression. Of course, Wendell knew that words have "significance," conferred on them by the arbitrary conventions of what he called "good use" (pp. 13, 17, 25); but what mainly attracted his fancy was a curious conflict that he saw between the conventions of good use and the writer's desire to achieve "effects" under the rule of "certain very simple principles of composition" (pp. 150–155, for example; for the importance of effect in Wendell's system, see further on, p. 22).

Wendell's description of "the experience of pretty much every writer" shows just how little he was concerned with the substance of expression. First, "[an] idea presents itself to [a writer] in a general form . . ." This "idea" may be either "some fact in experience . . . which nothing but the most exquisite verse can adequately formulate" or nothing more than "an invitation to dinner which he wants to accept." Then the writer begins his first and often longest task, that of planning the physical organization of his work, his piece of style: "how to begin, what course to follow, where to end." Finally the writer must "fill out his plan," must "compose in accordance with the general outline in his mind, a series of words or sentences which shall so symbolize this outline that other minds than his can perceive it" (p. 116). Did Wendell really mean to reduce comprehension to the perception of an outline? The outline of the acceptance of a dinner invitation?

The passage is remarkable for words that remove writing from any concerns of living and turn it into the construction of beginnings and endings, the filling out of plans, the narrowing and the simplification of material and analysis so that—for example, in *English Composition*—no chapter "should contain anything that could not ultimately be included in a general summary of no more than one paragraph" and indeed so that all eight chapters should "contain nothing that could not ultimately be summarized almost as compactly." He advised writing the summaries first and using "them as guides, rather than masters" (pp. 155–156). The influence of legal training on the theories of our early masters has yet to be noticed. At any rate, if students follow such a contrived, mechanized procedure as Wendell recommended, young people, learners are bound to produce just what they do, "papers" of generalities, thinly supported by largely imaginary illustrations or examples.

After that schoolmasterly advice, Wendell inserted a comment on Carlyle's *Frederick the Great* to illustrate unity, the first of his famous trio of "Principles of Composition" ("Note to Teachers," p. 2). Because it has not merely "a single definite title" but also "one central figure who gave his name to the whole," Wendell thought it—Aristotle notwithstanding—"a most notable example of unity" (pp. 157–158). He followed the Carlyle illustration with a comment on the well-known "trait" of English "men of genius" to write with "utter disregard of form" and gave the obligatory comparison of Shakespeare and genius with Jonson and art. Apparently, Shakespeare's carelessness with form is evidence that his plays lack unity, some of them not even having "single definite title[s]." Jonson is the man to follow (in spite of his occasionally rather expansive titles) because every detail in his plays "is part of a preconceived and complete whole" and because most people "are not great enough to disdain rule and principle and conscience"—the formal "conscience of the artist," that is (pp. 159–162 passim).

Wendell followed this miniature, two-part excursus with an elaboration of the unity principle:

> In planning our compositions, then, there is nothing quite so important as a constant, conscious determination that they shall contain what belongs there, and nothing else; that, if any work of ours can make them, they shall group themselves about one central idea,—that they shall have unity [p. 162].

I find some sort of careless phrasing in, on the one hand, "if any work of ours can make them" and, on the other, "they shall group themselves." One would expect "we shall group them" or something of the sort. And how do compositions "group themselves" or "be grouped" either? What in the compositions is the subject of "group"?

Wendell began his next paragraph with this: "So much for the substance of our whole composition" (p. 162). The "So much for" seems brusque. And I am not sure that Wendell's description of the material in *Frederick the Great*—"the mass of living facts" that Carlyle had "flung together" (p. 158)—is particularly convincing as evidence of his understanding of, or interest in, Carlyle's philosophy, which most people would suppose to be the point, if not the "substance," of Carlyle's work. Did Wendell think of *Frederick the Great* as just one massive collection of daily themes?

It is hard to find evidence that Wendell was at all interested in discursive, analytical thinking. He could say, "Before we can tell anything about form we must understand much about substance." But then, instead of discussing what must be known about substance or how it should be learned or invented, Wendell went into a rather poetic passage on the difficulties of thinking, of disentangling "from the riotous thicket of thought and emotion . . . the exact thoughts and emotions whose mutual relations as well as whose independent selves shall serve our purpose of imparting to readers what we have in mind . . ." (p. 136).

That view of mind is repeated a few pages on, in a description of "the normal condition of the human mind," in which ideas exist "in a state of confusion." "Dozens of trains of thought," he said, "are running in our heads at all times, intermingling, distorting one another, entangling themselves a great deal more than any one who does not sometimes try to disentangle them would suspect" (pp. 152–153). If we put aside the question how much of that may be based on introspection, it does at least seem pretty clear that Wendell thought that thoughts and emotions, his "substance of expression," exist prior to any particular situation or occasion and that composing consists merely in plucking them from the thicket (stream?) of consciousness. It is a curious conception of reasoning.

Perhaps in such passages Wendell was just cannily working in allusions to William James's novel descriptions of consciousness and perception, using James to give a modish philosophical foundation to his pedagogy. Did he mean it is because of the blooming, buzzing confusion of reality that writers must carefully consider their "material" and "pretty carefully [construct] the proper mould"? Apparently "material" means only "our thoughts and emotions." And "proper mould" seems to mean no more than "a form which shall make them [thoughts and emotions] intelligible to ourselves and others." So the whole elaborate and infelicitous advice comes down to no more than that anyone who "would cast anything into any form must first proceed to make a mould" (pp. 152, 153), or outline. As the figure in "cast" and "mould" implies, Wendell's dictum will work for, say, metal or certain kinds of glassware. But I don't think it serves the writing student very well. Nor does the figure get very close to the principle that some part of composing involves the refining or sophisticating of the notions that writing begins in or that writers begin with. I am thinking about the ideas that (using Wendell's word) "present" themselves to writers in more or less general form and that they then explore for meanings, for facts, data, observations, what not—the stuff out of which significances are made or in which they are discovered.

I suppose Wendell hit on that "cast"/"mould" figure because he had noticed—noticed in some fundamental way—"that written words address themselves to the eye and spoken words to the ear." In his mind this was a simple and basic "physical fact" that had been "neglected by the makers of textbooks." A person doing a composition—a sentence, a paragraph, a whole piece of style—must, therefore, think first of "what catches the eye," which is "obviously not the immaterial idea a word stands for, but the material symbol of the idea,—the actual black marks" on the page. In fairness, I must add that Wendell was not entirely oblivious to the very "subtile [his consistent spelling] and varied significance" that the little black marks had come to have attached to them (p. 32).

But he was first interested in the appearance on the page of the "outward and visible signs" of meaning (p. 69). And "what the words stand for" seems to enter his calculations only in some secondary way, "when we come to consider the substance of a composition" (p. 29). And considering the substance of a composition was not, for Wendell, to test or examine ideas and material, but rather to determine "what ideas we wish to group together" (pp. 29–30). Indeed, for Wendell "the iron utensil frequently employed for purposes of excavation" is a more complex idea than "spade" (p. 41, repeated at p. 63).

I have sometimes thought that the doctrines and dogmas of *English Composition* can all be explained by the pedagogical need to simplify, to develop a composing process that will produce themes that can be given "a hasty categorical analysis" rather than be the means whereby young people can use their language toward growth in their thought. With *English Composition* the effect is much more one of constriction than of growth in thought. This can be vividly seen in Wendell's test for unity in a paragraph, which was to construct "a sentence whose subject shall be a summary of [the paragraph's] opening sentence, and whose predicate shall be a summary of its closing sentence" (p. 129). To illustrate this "matter so technical," Wendell worked over a four-paragraph editorial from *The Nation* that he had been using in his teaching since November 1889. I give his summary of the first paragraph and his "general summary" of the whole editorial:

> The decline in the proportion of [college] students to the population . . . is noticeable in the United States and England.

> The decline in the proportion of students to population . . . makes "Study or clear out" the proper motto for any college [pp. 129, 130].

The first parts of those sentences, by the way, are the subject of the first sentence in the editorial; the second parts are the predicates of the last sentences of the first and last paragraphs of the editorial. It is hard to think what benefit students or his audience would have got from watching Wendell work through that tricky procedure.

The system of *English Composition* reduces composing to consideration of abstractions like Mass, the second of Wendell's principles of composition. (Coherence is the third.) Mass is "the principle which governs the outward form of every composition." In keeping with "outward form," which itself follows from Wendell's theory of the connection between the written word and the eye (p. 32), Mass is treated as a matter of sight. Expanded, the principle is "that the chief parts of every composition should be so placed as readily to catch the eye" (pp. 32, 127). But Wendell went further. As if citing a fact derived from observation or even experiment, he declared, "Trained or untrained, the human eye cannot help dwelling instinctively a little longer on the beginnings and ends of paragraphs than on any other [physical] points in the discourse." For evidence he went to the experience of book reviewers and of lawyers trying to collect authorities for a brief, surely two rather special activities.

Such people skim their reading, operating "quite independent of sentence-structure, and of unity, and of coherence," intending simply to discover the "visible, external outline" of the piece of style (pp. 127–128). Skimming "means," Wendell said, "that the beginning and the end of a paragraph are beyond doubt the fittest places for its chief ideas, and so for its chief words" (pp. 127–128, slightly modified at p. 180). I do not know how a paragraph with unity can be said to have "chief ideas." But no matter, the real problem is that, once again, Wendell was directing attention to the appearance of the words on the page, their simple placement, and away from the value of argument, the validity of evidence, the warrantability of conclusions. Just as skimming in reading is likely to lead readers to ignore nuances and qualifications and to frame generalities that are at least inexact, so concentrating on plan and appearance is likely to produce what Wendell called, more aptly than he knew, "pieces of style" (p. 7).

For Wendell, the writer's "real question . . . is what effect he writes to produce" (p. 67). In general, he said, "writers wish to produce an effect of firm precision"; so they write "paragraphs that have coherent unity and a firm mass." But anyone who wants to suggest confusion or indecision can do so, in the first case, "by deliberately disregarding coherent unity," and in the second, "by deliberately weakening of mass" (pp. 146, 147, also 190). Of course, writers are interested in the abstract qualities of their writing, and, of course, they think about their words and the movement of their sentences. But most of them try to think also about what they're saying. That is a subject that does not get much attention in *English Composition*.

V

Evidently Wendell saw himself doing a quite different job from ordinary teachers and textbook makers. Not for him the rules of rhetoric though he seems to have been content with the prevailing classification of nonconforming usages into Barbarisms, Improprieties, and Solecisms, and he even had his students searching for examples in the papers of their fellows (Note for Teachers, pp. 1, 2, 44–50, 78, 81). Wendell sought to teach students to develop their own senses of effect and to learn to capture the moment, enforcing their meanings in the process (p. 147) by choosing among the various molds in which sentences and paragraphs may be cast, within the limits set by Good Use (pp. 150–151, 120).

It sounds very liberating indeed, especially if it has added to it Wendell's even loftier summation:

> If teacher or pupil keep himself down to the symbol alone [words and groups of words], he sinks hopelessly into the depths of pedantry. But if teacher or pupil keep himself alive to the truth that what he is striving to accomplish is no less a thing than an act of creative imagination; if he learn to know that in his own little way he is trying to do just such a thing as the

greatest of the masters have done before him; if through the symbol his eye learn to seek and know the infinite variety of truth that lies behind,— he will find that even though technical mastery never come, he will learn more and more the infinite, mysterious significance of that human life that each of us is living for himself [pp. 306–307].

Devising a title for his book was hard, Wendell found. He started with "Rhetoric," but rejected it as too closely connected to "the art of persuasion" and to "the art of polite embellishment of language." Then he tried and rejected in succession "Style," "The Philosophy of Style," "Art of Composition," and "Literary Composition." (The titles do show how his mind was running.) Finally, he "fell back at last on the not very enticing title, 'English Composition,' which is used at Harvard College to describe the subject with which the lectures dealt." At least, he thought, that title was "unmistakably specific" (pp. 214–215).

There is something depressing about that descent from "Rhetoric" to "English Composition," even considering what rhetoric had been reduced to by Wendell's day. And it is easy to sympathize with the feeling behind Wendell's somewhat fulsome explanation of his aims for the dailies.

What I bid them chiefly try for is that each record shall tell something that makes the day on which it is made differ from the day before. Dreadfully dull work they think this will be, and dreadfully dull most of my friends think the task must be of whoever reads these records. Yet as the college year goes on, the task generally grows less and less dull to the writers; and to the readers it is generally far from dull. Each new bundle of these daily notes that I take up proves a fresh whiff of human life, as day by day it has presented itself to real human beings . . . [pp. 265–266].

But so far as Wendell was able to say, the dailies led the boys no further than to "begin to find out for themselves how far from monotonous a thing even the routine of college life may be, if you will only use your eyes to see, and your ears to hear"; and "to feel . . . that this real human life of theirs, this human life that is peculiarly theirs, is the source from which they must draw whatever they really have to say" (p. 266).

VI

According to Wendell, I should now be able to sum up all in a paragraph. But to sum up Wendell and his work were (to use one of his subjunctives) to attempt the impossible. So I shall simply borrow a bit of verse gently mocking Wendell's values, his style, in a way, I suppose, his very being:

Please make a careful study of this truthful illustration
And take especial notice of the subtle connotation.
The atmosphere of London is so well suggested there,
You'd think you were in "Rotten Row" instead of Harvard Square.
How palpably inadequate my feeble talents are
To tell what Harvard culture owes to this, its guiding star!
Coherence, mass, and unity in Barrett are combined,
To edify the vulgar, and abash the unrefined.

The lines are by Henry Ware Eliot, Jr., of the Class of 1902. They appeared in *Harvard Celebrities,* a book of caricatures and mildly satirical verse that he and two other undergraduates published in 1901. I think Ware's lines capture most of what I see as the essence of Wendell's human and pedagogical characters. I can add to them this memory by another student, which J. Donald Adams, not likely to have been an unfriendly reporter, included in his biography of Charles Townsend Copeland, another of the Great Teachers around Wendell during the Golden Age. The student recalled "that Wendell was pleased whenever someone turned in a theme on a broad subject like the mores of those ladies of an ancient profession who adorned the houses on Bulfinch Street. Compositions like these were likely to receive an A" (p. 98).

I conclude with that detail, not to emphasize Wendell's eccentricity, his interest in the bawdy, but rather to suggest just how close to the surface of subjects Wendell's students must have been able to keep their themes. How much could they have said about "those ladies of an ancient profession who adorned the houses on Bulfinch Street"? Adams's language is answer enough.

References

Adams, J. Donald. *Copey of Harvard.* Boston: Houghton Mifflin, 1960.

Brown, Rollo Walter. *Harvard Yard in the Golden Age.* New York: Current Books, 1948.

Engle, Lars. "The Remarkable Course Called Daily Themes." *Yale Alumni Magazine and Journal,* April 1983, 16–19.

Howe, M. A. De Wolfe. *Barrett Wendell and His Letters.* Boston: Atlantic Monthly Press, 1924.

James, Henry. *Charles W. Eliot.* 2 vols. Boston: Houghton Mifflin, 1930.

Morison, Samuel Eliot. *Three Centuries of Harvard.* Cambridge, Mass.: Harvard University Press, 1942.

Santayana, George. *The Middle Span,* vol. 2 of *Persons and Places.* New York: Scribner's, 1945.

Self, Robert T. *Barrett Wendell.* [Boston]: Twayne Publishers/G. K. Hall, 1975.

Weeks, Edward. *The Lowells and Their Institute.* Boston: Little, Brown/Atlantic Monthly Press, 1966.

Wendell, Barrett. *English Composition.* New York: Scribner's, 1891, 1899.

Selected Publications of Barrett Wendell

BOOKS

Cotton Mather, The Puritan Priest. Makers of America Series. New York: Dodd, Mead, 1891.

English Composition. New York: Scribner's, 1891.

Stelligeri, and Other Essays Concerning America. New York: Scribner's, 1893.

William Shakespere: A Study in Elizabethan Literature. New York: Scribner's, 1894.

A Literary History of America. The Library of Literary History. New York: Scribner's, 1900.

Liberty, Union, and Democracy, the National Ideals of America. New York: Scribner's, 1906.

The France of Today. New York: Scribner's, 1907.

The Privileged Classes. New York: Scribner's, 1908.

The Mystery of Education, and Other Academic Performances. New York: Scribner's, 1909.

ARTICLES

"Mr. Lowell as a Teacher." *Scribner's Magazine* 10 (November 1891), 645–649.

"Some Neglected Characteristics of the New England Puritans." *Annual Report of the American Historical Association for 1891.* Washington, D.C.: American History Association, 1892.

"English Work in the Secondary Schools." *School Review* (1893), 638.

"English at Harvard." *The Dial 16* (March 1894), 131–133.

"Samuel Eliot." *Proceedings of the American Academy of Arts and Sciences* 34 (May, 1899), 646–651.

"The Privileged Classes." *Journal of Education,* February 27, 1908, pp. 11–24.

"Thomas Raynesford Lounsbury (1838–1915)." *Proceedings of the American Academy of Arts and Sciences* 53 (September 1918), 831–840.

"James Russell Lowell." *Commemoration of the Centenary of the Birth of James Russell Lowell.* New York: Scribner's, 1919.

FRED NEWTON SCOTT

by Donald C. Stewart

On 13 November 1916, Gertrude Buck, one of Fred Newton Scott's earliest and most gifted students in rhetoric at the University of Michigan, wrote to Georgia Jackson, telling her that "it seems to me that Professor Scott stands for contributions to rhetorical theory far more distinctively than for any teaching of journalism. Other people have done the latter but he is almost alone in the former field. This is what his reputation will ultimately stand on, I am convinced, and a fund for publication in this line would recognize the fact as it ought, I believe, to be recognized."[1]

It has taken nearly three-quarters of a century for Buck's prophecy to be tested, but there are understandable reasons for this fact. After Scott died in 1931, he was quickly forgotten by the profession he had served so long and ably, primarily because it had abandoned rhetoric and turned its attention almost entirely to literary scholarship. Even more to the point, Scott was a man of such wide-ranging interests and competencies that it would have been difficult even for his contemporaries to decide exactly what his intellectual legacy would be. Half a century away from his death, we have, at last, a much clearer perception of that legacy although it has taken some time for it to come into focus. It is broader than Buck imagined, but she was ultimately correct in perceiving that Scott's contributions to rhetorical theory were his distinctive legacy. However, we must also note that these theoretical ideas were closely linked to his perception of the teaching of writing and the priorities that he felt were central to the discipline of English.

Scott was born 20 August 1860, in Terre Haute, Indiana, the fifth of Harvey David and Mary Bannister Scott's six children, but one of only three who survived into maturity. His father, a lawyer, had a long record of public service as both representative and senator in the Indiana legislature, as an Indiana representative in the thirty-fourth Congress (1855–57), and as judge of the Superior Court in Terre Haute. Scott was apparently educated at home until the Indiana Normal School (now Indiana State University) opened in 1870. Too young for college, he attended the college's training school, where he says he was introduced to the science of psychology by William A. Jones, the first president of the university.[2]

He remained in Terre Haute until late 1878, then moved to Battle Creek, Michigan, to become secretary to Dr. J. H. Kellogg of the Battle Creek Sanitarium, a brother to the Kellogg who founded the breakfast cereal company. Scott's purpose in making this move was most likely twofold: (1) to earn a Battle Creek diploma, which would gain him admittance without entrance examinations to the University of Michigan; (2) to earn money to defray his college expenses. He succeeded in both endeavors and matriculated at the University of Michigan in the fall of 1880.[3]

The progress of his education and the development of his professional career are much clearer after this point. He completed his B.A. at Michigan in 1884, his M.A. in 1888, and his Ph.D. in 1889. That year he joined the Michigan staff as an instructor in English. He was promoted to assistant professor in 1890, to junior (associate) in 1896, and to professor in 1901, by which time he had become a celebrity. According to W. R. Humphreys, "The man who during the years around 1900 attracted students from all over the country, and, notably, advanced students from the East, was Fred Newton Scott. In recognition of his eminence the Department of English was divided, and Professor Scott was for the rest of his time on the faculty head of the Department of Rhetoric."[4]

Humphreys was right about Scott's eminence but only partially right about the formation of the Department of Rhetoric, the most significant administrative action that involved Scott in his long career at Michigan. The Department of Rhetoric was created, not only because of Scott's reputation, but for an eminently practical reason. Isaac Demmon, head of Michigan's English Department, sent the Michigan regents a note in April of 1903, the substance of which was that because of increasing enrollments in literature classes, the English Department should be divided; that he head up the work in literature and that Scott, who had demonstrated long service and special competence in the area, be given charge of the work in composition and rhetoric.[5] One may speculate that Demmon was rationalizing his jealousy of a man who, because of *Paragraph Writing* alone and his visibility in the Modern Language Association, had achieved significant national recognition, but I am now hesitant to do that. Scott was clearly pleased to have the headship of a Department of Rhetoric, and his correspondence and daybook entries for years afterward indicated continued cordiality with Demmon.

As Humphreys indicated, Scott was, by this time, distinguished far beyond the borders of the Michigan campus. From 1896 to 1903 he was president of the Pedagogical Section of the Modern Language Association (MLA), and for a brief period at the turn of the century he attempted, through that division's programs, to generate discussion in the profession of questions relating to the nature and efficacy of rhetoric in the college curriculum. In 1907 he became MLA's twenty-fourth president; from 1911 to 1913 he served as the first president of the National Council of Teachers of English, the only one ever to serve two terms in that office. In the years 1913–14 he was president of the North Central Association of Colleges and Secondary Schools, and in 1917 he became president of the American Association of Teachers of Journalism. Scott also belonged to a number of literary clubs and professional societies, locally, nationally, and internationally; and he was in constant demand as a speaker, primarily before teachers' groups, all across the country.[6] His correspondence is full of letters negotiating his visits from New England to California.

Teaching, of course, is a much harder matter to judge, but the record here is also clear. John Dewey, in a sketch of Scott's early career, says that he took a course that students had to come to expect as extremely boring and gave it life and significance for them. Just nine months before he died, Jean Paul Slusser, Professor Emeritus in Art at Michigan, told me that Scott was a major influence in his life. Scott had advised Slusser to study abroad, which he proceeded to do. As a result of the experience, Slusser discovered his aptitude and passion for art, which eventually became his vocation. "We were all his devotees," Mr. Slusser said. He had had the privilege of sitting in Scott's classes when Scott was still in his middle years and full of the vigor that characterized his work at that time. Apparently, early in his career, Scott saw the weaknesses in a classroom in which students sat in rows, looking at each other's backs. His students seated themselves around a large oak table, which can still be seen in the Hopwood Room at the University of Michigan. Ray Stannard Baker, author of the first significant biography of Woodrow Wilson and a successful professional writer, gives this account of what happened to him as a result of his exposure to Scott and his teaching methods:

> Although he probably knows nothing about it, I have always felt that Professor Scott gave me more than any other teacher, at Michigan or elsewhere. I took his courses at a time when I was confused in my own mind as to my own capacities, as to what I could best devote myself to. I

found in the two courses with Professor Scott, which I deserted the Law School to take, the liberation which I was seeking. He seemed to have an extraordinary gift of setting men to thinking and then, by deft touches of advice, not too much of it, indicating the reading which would best enlarge the vista.

As a result, I incontinently deserted the law courses which I was supposed to be taking, and spent all my time at the library, developing themes which were suggested by the discussions in Professor Scott's seminars. I am sure he never knew the number of books I read, nor the darkness of intelligence I applied to some of them, but it resulted in the definite choice of my profession as a writer.

I think also, if it had not been for Professor Scott, I should have graduated, with the perfectly innocuous degree of Bachelor of Law, from the University of Michigan. But he made me so hot to begin the actual practice of the calling to which he inspired me, that I deserted the University, having learned how to educate myself, and never got the degree!—which was one of the happiest incidents of my life.[7]

Scott's publications in rhetoric and composition alone were no less distinguished than his service work and his teaching, but that has not been generally well known until quite recently.[8] One can, I believe, get a fairly clear perspective on them by breaking them down into three categories: (1) textbooks, (2) academic articles on pedagogy and the state of the profession, and (3) rhetorical theory.

Scott's textbooks are the least innovative of his work, but they are superior to the books of his contemporaries, such as A. S. Hill of Harvard and John Genung of Amherst, for two reasons: (1) They constantly stress writing in a social context, and (2) they incorporate theory that was in advance of its time. Scott's texts were numerous, and they provided him income that made his other professional activities possible. His first text—and one of the most enduring—was *Paragraph Writing* (1893), which he did in collaboration with Joseph Denney of Ohio State, a former classmate and colleague at Michigan. This book had its origins in a very practical problem teachers of the early 1890s were facing: a Report of the Committee on Composition to the Harvard Board of Overseers in 1892 had suggested that high schools were to blame for graduating students who simply were not literate. A spate of articles in the popular press was generated by this report, and a back-to-basics movement of the time was born. Like that of the mid-1970s, however, it focused too much on superficial mechanical correctness (spelling, punctuation, and usage). As a consequence, teachers in the schools were under pressure to require more writing, and the number of papers they were faced with grading was, to say the least, a Herculean labor. Scott and Denney offered them a way out of the dilemma, one that was thoroughly grounded in the pedagogy of the time. *Paragraph Writing* is essentially an argument for teaching the paragraph

as a theme in miniature. It grew out of a pamphlet that Scott and
Denney had written in 1891, then quickly developed into a book that,
according to the authors, "aim[s] to make the paragraph the basis of a
method of composition, to present all the important facts of rhetoric in
their application to the paragraph."[9]

The authors cite, as sources for their conceptions about the paragraph, Alexander Bain, A. S. Hill, D. J. Hill, J. S. Clark, T. W. Hunt, G.
R. Carpenter, J. G. R. McElroy, John Genung, and Barrett Wendell.[10]
Their separation of discourse into the sentence, the paragraph, and the
essay is much like Wendell's word, sentence, and paragraph categories.
Their insistence that all paragraphs should exhibit unity, selection, proportion, sequence, and variety is an adaptation of Bain's rules for paragraphs with echoes of Wendell's famous triad of Unity, Coherence, and
Mass. Their justification of the paragraph as the ideal unit of discourse
to study, for both practical and developmental reasons, seems, however
to be original with them. They cite the work of Scott's sister, Harriet, in
the Detroit Training School for Teachers (published in the Report of
the Detroit Normal Training School for 1893) as evidence that students
comprehend a paragraph more readily than they do single sentences.
The sentence, they argue, is "too simple and fragmentary" to be
adapted, pedagogically and psychologically, to the needs of the classroom. The entire essay is too long and too complex. The paragraph,
the intermediate unit of discourse, is both long enough to give students
a sense of the principles on which units of discourse are formed and
short enough for them to comprehend. In addition, teachers can assign
many more paragraphs than they can entire themes, with the resulting
relief from paper grading and benefits to students from increased writing practice. Study of the paragraph is, furthermore, a natural introduction to the study of longer units of discourse.

Scott and Denney were not through, at this point, with their treatment of the paragraph. In the third edition they attempted to develop
a theory of the paragraph that was grounded in the psychology of
composing. Rejecting the notion that thought proceeds in a smooth,
unruffled, and orderly way, they say that

> the thought-process consists of a series of leaps and pauses. The stream
> shoots toward some point of interest, eddies about it a moment, then
> hurries on to another. "In all our voluntary thinking," says Professor
> James (*Psychology*, I, 259), "there is some topic or subject about which all
> the members of the thought revolve. Half the time this topic is a problem,
> a gap we cannot yet fill with a definite picture, word, or phrase, but which
> influences us in an intensely active and determinate psychic way. Whatever
> may be the images and phrases that pass before us, we feel their relation to
> this aching gap. To fill it up is our thought's destiny." Toward this objec-
> tive point the thought presses with an imperiousness that is no inadequate

test of the value of the process. The feeble mind feels only in a vague way the propulsion toward the central idea; the genius often flies toward the goal as unerringly as the armature leaps to the magnet [pp. 94, 95].

The most noteworthy feature of this pedagogy, even if we do not agree with it today, is that it represented some attempt to ground a theory of the teaching of writing in psychology. Although I cannot, at this time, report accurately on the full extent of his debt to William James or John Dewey, for example, I am quite sure that Scott was alone in his time in recognizing how significantly other disciplines could enrich the study of composition and rhetoric.

I want to conclude this brief examination of what may have been his most famous textbook with one other observation. At the end of the section on paragraph theory, Scott and Denney offer a very brief bibliographical essay, in which they cite treatments of the paragraph that have influenced their thinking. And they note the only significant scholarly study of the paragraph in that era, Edwin Lewis's *The History of the English Paragraph*, a doctoral dissertation that was subsequently published by the University of Chicago Press in 1894. My general point is that here we have a textbook, drawn not only from the authors' immediate practical experience of teaching, but thoroughly grounded in good scholarship and enriched theoretically by insights from another discipline.

Scott's other texts, usually done in collaboration with others, range over a variety of topics and are directed at a broad spectrum of students. With Denney, he collaborated on *Composition-Rhetoric* (1897; rev. in 1911), *Elementary English Composition* (1900; rev. in 1908), and *Composition Literature* (1902). The *Composition Literature* text was revived by Edward P. J. Corbett and Virginia Burke as *The New Century Composition-Rhetoric* (New York; Appleton-Century-Crofts, 1971). Scott also collaborated with Gertrude Buck on *A Brief English Grammar* (1905) and with Gordon Southworth in *Lessons in English*, Books I and II (1906; rev. in 1916). The latter was essentially a high school text.[11]

The collaborations with Denney contain, for that time, a fairly conventional listing of topics to be treated (the word, the sentence, and the paragraph) and the forms or modes of discourse (narration, description, exposition, argument, and poetry).[12] However, all these books contain occasional observations that reveal both theoretical and philosophical concerns that were small oases in the arid wastes of composition instruction at that time. For example, in the opening chapter of the *Composition-Literature* text, they lay out some first principles whose common sense is unique for that time simply because common sense did not inform the work of many teachers of writing then or even—one might be tempted to say—today. If you want to know how anyone who is good at anything acquires that proficiency, Scott and Denney say, you ask the people who are experts in the field. Thus, students of writing should get their instruction, either directly or indirectly, from successful writers of essays, novels, poems, and so on. They then quote from Frederick W. H. Myers, English essayist and poet, who says that "almost the only way to write effectively is to choose some subject on which one really feels deeply and has thought long, and then to select and arrange one's language with a strong desire that one's readers shall understand just what one means, and be persuaded that it is true."[13] Scott and Denney extend this remark by saying that "one who would write well must be deeply and sincerely interested in his subject. He must see things with his own eyes, think about them with his own mind, have his own sincere feelings about them, not pretend to the feelings of some one else. Then his words will have an honest ring, and will be in some measure alluring to others" (p. 9).

The *Composition-Rhetoric* is informed by three principles, two of them sound but not particularly new, and a third, which does break new ground in the pedagogical theory of the time. Principle one was the merging of rhetorical theory and practice (apparently they were reacting against teaching rhetorical theory as a body of knowledge separate from its application in composition), and principle two was the use of the paragraph as a systematic method of instruction. Principle three is of particular interest to us today, however, in that it is an early focusing of attention on the process, not the product, of writing.

> A third idea which informs the work is the idea of growth. A composition is regarded not as a dead form, to be analyzed into its component parts, but as a living product of an active, creative mind. The paragraph is compared to a plant, springing up in the soil of the mind from a germinal idea, and in the course of its development assuming naturally a variety of forms. This kinetic conception of discourse, besides being psychologically more correct, has proved to be practically more helpful and inspiring in composition classes than the static conception which it is intended to displace. Where it has been employed, pupils attempt various forms of self-expression with greater willingness and confidence, and their efforts are attended with greater success [p. iv].[14]

Always attentive to the attitudes and needs of students, Scott and Denney stress, in *The New Composition-Rhetoric*, a revision of the earlier work, that "composition is regarded as a social act, and the student is therefore constantly led to think of himself as writing or speaking for a specified audience. Thus not mere expression but communication as well is made the business of composition" (p. iii).

This particular idea is developed further in the authors' preface to *Elementary English Composition*. They list three causes for the indifference of students taking English composition: (1) Too often the work in such classes is mere repetition of what students have learned before, "affording them no new view of their English, and calling for the exercise of no new form of ingenuity that might enlist their interest" (p. iii); (2) students are not taught the social consequences of expressing themselves; (3) they are too often confronted with the separation of written and spoken discourse.

> The artificial separation of two things which naturally belong together takes the heart out of both of them. Hence we find in the schools writing that is feeble and impersonal, and oratory that is flamboyant and insincere. That the simple utterances of daily desires and needs are as truly compositions as the most labored essays, that essays are best when they are the simple utterance of daily desires and needs, are lessons which pupils, if they have not already learned them, cannot learn too early in their secondary education [p. iv].

This is a lesson teachers, bent on pushing students into academic discourse before they are ready for it, cannot learn often enough.

Although Scott and Denny's methods are not particularly original in the textbooks, these philosophical positions dictate the contexts in which the modes of discourse and matters of style and editing take place. For example, Scott was a believer in the use of pictures to stimulate writing, and his office-classroom in Old West Hall was ringed by a great variety of prints. And even in matters of editing, copyreading, and neatness, these texts display unusual common sense. Students are told not to learn conventions for their own sake; they are told to learn them to avoid confusing readers. One of the authors' most humorous and effective examples comes in a discussion of the value of readable penmanship. Horace Greeley, as the story goes, responded to a lecture invitation from a club president in Illinois as follows: "Dear Sir: I am overworked and growing old. I shall be sixty next February 3d. On the whole, it seems I must decline to lecture henceforth, except in this immediate vicinity, if I do at all. I cannot promise to visit Illinois on that errand—certainly now now." Unfortunately, Mr. Greeley's penmanship was terrible, and he received the following astonishing reply: "Dear Sir: Your acceptance to lecture before our association next winter came to hand this morning. Your penmanship not being the plainest, it took some time to translate it; but we succeeded, and say, your time, February 3d, and terms, $60, are entirely satisfactory. As you suggest, we may be able to get you other engagements in the immediate vicinity; if so we will advise you."

The collaboration with Gertrude Buck, *A Brief English Grammar*, is important today as an example of Scott's enlightened and liberal attitudes toward usage, language in a social context, and the relationship of grammar study to composition. I do not wish to minimize Buck's contribution to this effort. I wish only to underscore the fact that on many of these issues she and her graduate professor spoke with one voice. For example, they note in the preface to the book that

> it seems the invariable tendency of any complicated system of forms, when made a subject of study, to cut itself off from the living processes which gave rise to it, and become in the student's mind mere matter, an arbitrary thing-in-itself, dead and meaningless. The danger of this tendency has been abundantly recognized in recent textbooks of English grammar and composition, which have attempted by various methods to recall the meaning to the form, to re-connect the word on the page with the thought which created it and the situation which shaped and modified it [pp. iii, iv].

They go on to say that the fundamental postulates of their book, the social function and the organic structure of language, have long been familiar to advanced students of philosophy and linguistics. These themes, as we have seen, emerged in Scott's other textbooks. However, another issue, one with which Scott would not have been so comfortable, surfaces in this text:

> Since the sentence, whether simple or elaborate, represents the typical act of thought-communication from speaker to hearer, the business of grammar is to define successively every member of the sentence, every word and every inflection, on the basis of its individual contribution to this act of communication. Thus the sentence is both the beginning and the end of grammatical study; and sentence-analysis, in the largest sense of the term, is its entire subject-matter [pp. 3, 4].

This assertion is very much in line with contemporary theory, which says that sentence competency is the basis of literacy. However, Scott and Denney's work with paragraph theory, their assertion that the paragraph was unique as a piece of discourse, more satisfactory than the sentence or the extended essay as a unit for study, may have made Scott uncomfortable with the assertion that the sentence was the fundamental unit of grammatical study. (Entries in his daybook at this time reveal that something about this collaboration was causing him distress, but he is never specific on the reasons for his discomfiture.) The problem arises because Scott, like modern theorists, brings rhetoric, linguistics, and psychology to bear on the fundamental question of literacy. The pedagogical issue resolves itself into this question: Is sentence competency, once achieved, the door to mastery of larger and more complex units of discourse, or should we focus on the paragraph and, becoming literate in that, achieve both competency in producing sentences and extended units of discourse? I do not believe the question has yet been satisfactorily answered.

There are other ideas in this text that, although held by a relative few in their time, became staples of linguistic theory later in the century: (1)

that the written language has its basis in the spoken language; (2) that the child has a functional mastery of the basic forms of English, acquired from parents and friends, long before he or she comes to school; (3) that grammar is descriptive of the speaking and writing habits of the users of a language, not something prescribed before the language was invented.

> It is sometimes said that grammar is a collection of rules for correct speaking and writing; but this is not strictly true. The rules of grammar are, like the laws of any physical science, such as chemistry, physics, astronomy, or physical geography. These sciences are not a collection of rules telling the winds and tides, for instance, what they must do, or prescribing how a certain acid or a certain base shall unite. They only report and explain what happens. And so grammar does not say to us directly, "You must speak thus and so," but only "English people at the present time do speak thus and so, for the following reasons." Knowing this fact, we shall undoubtedly choose to speak so, too, in order to be easily and precisely understood by others, or to avoid the appearance of eccentricity or ignorance; but grammar goes no further than to report and explain the usages of the language, leaving us to choose for ourselves whether we shall follow them or not [pp. 12, 13].

The *Lessons in English,* a collaboration initiated by publisher Benjamin Sanborn between Scott and Gordon Southworth of the Somerville, Massachusetts, schools, does not contain new material. Most of what is here is fairly traditional in literature, composition, and grammar, but, even on this occasion, Scott did not miss his chance to reform teachers who were doing English, as a subject of study, more harm than good:

> Too many teachers think of a text-book as a kind of machine-gun, built to fire with deadly precision so many loads a minute. This is a vicious error. A text-book should be the teacher's friend, guide, and helper. It may be a powerful aid and resource; it can never take the place of the teacher's personal enthusiasm, sympathy, and stimulus [pp. vii, viii].

Scott and Southworth conclude their preface with these observations:

> A final word of caution may not be out of place. Teachers of English are apt to attach much importance to the formal side of their instruction and to assume that a pupil's facility in reciting rules and detecting errors of speech is a sure sign of progress. The formal side must not, of course, be overlooked, but it should never be forgotten that the end of all instruction in English is growth in power of expression and appreciation. Drill which contributes to this end is good. Drill which, falling short of this end, merely fills the child's mind with rules and symbols, is a grievous waste of time [p. viii].

Over the course of his long career, Scott spoke out on a number of issues affecting the profession. Three of them surface again and again in his thinking and writing: (1) his passionate commitment to the teaching of English; (2) his skepticism about the efficacy of college entrance requirements; (3) the relevance of instruction in English in his time.

Scott's commitment to the teaching of English is most strongly stated in the chapters on the preparation of teachers that he wrote for *The Teaching of English in the Elementary and Secondary Schools,* a collaboration with George R. Carpenter and Franklin Baker, both of Columbia University. In general, Scott says, the English teacher should be able to speak and write clearly and without affectation, should be well read in English literature, and should have some competence in foreign languages. The English teacher's special skills should include the ability to read and correct themes efficiently and with maximum profit to the student (possible *only* if the teacher has a thorough grounding in rhetorical theory and history), knowledge of the history of the language and grammatical theory without which he or she cannot teach grammar effectively, and a knowledge of literary criticism because the latter informs the teacher's work in the other areas. Everything Scott says here is as relevant today as it was in 1908. The entire section is a treasure trove of good advice for anyone preparing to teach English at any level. But underlying all Scott's good advice about the knowledge and competencies that the English teacher must acquire is a philosophical attitude that, for him, was essential to anyone aspiring to enter our discipline:

> The teacher who has not a passion and an aptitude for imparting instruction in English, who does not feel that it is the great thing in life to live for, and a thing, if necessary, to die for, who does not realize at every moment of his classroom work that he is performing the special function for which he was foreordained from the foundation of the world—such a teacher cannot profit greatly by any course of training, however ingeniously devised or however thoroughly applied [pp. 307, 308].

Entrance requirements in English had become a subject of concern to the profession during the late nineteenth century because there was a rapid increase in the number of students going to college, and many were not prepared to write as well as they should have been. College faculties, accustomed to receiving the relatively well-educated sons of gentlemen, sought ways of screening young men of different interests and abilities and markedly uneven preparation in English. The creation of entrance examinations was one response. (Others were the creation of remedial courses, preparatory programs, even new colleges.) Teachers' attention on these examinations, as one might suspect, was caught most immediately by errors in spelling, punctuation, and usage. The Harvard experience is illuminating. Harvard began requiring entrance examinations in English in 1874, and from the beginning the examiners, despite their protestations of concern with substantive matters like content, structure, and style, were preoccupied by superficial mechanical errors more than anything else. The most acerbic comments about errors in students' performances on the entrance examinations came from Adams Sherman Hill, director of the composition program at Harvard from 1876, when he became the fifth Boylston Professor of Rhetoric, until his retirement in 1904.[15]

One of the reasons Scott did not like entrance requirements was that they reflected superficial notions about the nature of English as a discipline. In "The Report on College Entrance Requirements in English," *Educational Review,* October 1900, he listed a number of issues a committee charged with studying the problem failed to address:

> [Teachers] had a right to expect that it [the committee's report] would . . . summarize the history of the teaching of English for the past generation; that it would classify and critically review the various methods that now compete for popular favor; that it would give the results of experiments in the teaching of literature and composition; that it would discuss the special training of teachers; that it would discriminate between elementary and advanced work in methods, in choice of subjects, and in the character of recitation; that it would summarize and review the best of the literature on English teaching that has appeared of late; that it would treat of the methods and devices by which the labor of teaching and especially of essay-correcting may be lightened; and finally, that it would make plain in what respect this report is conceived to be an advance upon its predecessor, the report of the committee of ten [pp. 291, 292].

It is apparent that Scott was really galled here by a shallow perception of a very complex subject. "I think I speak for the majority . . . when I say that the most characteristic thing about English teaching at the present time [1900] is its unsettledness. It is fuller of unsolved problems than any other subject that can be mentioned. It is a kind of pedagogical porcupine" (p. 292). He follows this with one of the most incisive and fundamental criticisms of the profession, its priorities and methods, that one will find anywhere in his work:

> *Are* our methods of instruction in English in harmony with the social demands of our great industrial community? I suspect that they are not. More than that I suspect that the hard knot of the English question lies right here—that our present ideals and methods of instruction are in large part remnants of an adaptation to a state of things which long since passed away [pp. 293, 294].

A year later, in "College Entrance Requirements in English," *School Review,* June 1901, he expressed himself on the philosophy of entrance requirements, which he regarded as expressions of some fundamental differences between educators on the essential nature of a university. He contrasted a feudal conception, which sets admittance standards and lets students in only if they qualify (and thus exists in isolation from the rest of the world), with an organic conception, which sees the university as an integral part of an entire interdependent learning organization (including the elementary and secondary schools), which functions most efficiently when all parts work cooperatively together. Specifically, he was contrasting schools like Harvard (feudal) with Michigan (organic), and he did not disguise which conception he thought more beneficial to the society it served.

How do these divergent conceptions of a university affect entrance requirements in English? Scott says that the feudal system encourages arbitrary requirements, rigid conceptions of what good writing should be, and teaching that becomes coaching for passing entrance exams. "I can think of no better recipe than that [the feudal system] for deadening the nerves of sympathy and enthusiasm" (p. 371).

The organic conception answers the teacher's questions about entrance requirements thus:

> We make no formal requirements. We only point to our needs. What we want is young men and women whose literary instincts are normal and whose literary habits are good. We want students who know what good literature is and enjoy reading it; who can express themselves with a fair degree of ease and accuracy; and who have a taste for what is simple and sincere, as opposed to what is tawdry, or mawkish, or vulgar, in their own writing and the writing of their fellow students [p. 373].

Scott drew on the opposition between feudal and organic conceptions of the university for another blow delivered at the mentality that spawned entrance examinations and used them as measures of student intelligence and competence. In "What the West Wants in Preparatory English" (1909), he says that the feudal conception has a baleful effect on teachers, students, and the course, turning the latter to dry rot. It is a system, he believed, calculated to make all involved hate a subject that should be enriching and enjoyable. He saw this system as the source of the comp-lit course,

> . . . an unfortunate entanglement . . . of literature and composition. That any great gain comes to the student either in literary appreciation or in the command of his mother-tongue from the incessant writing of outlines of plots, critical estimates which ape maturity, or characterless sketches of character, has not, I believe, been demonstrated. On the other hand, it is the experience of most teachers with whom I have discussed the question, that such essays, especially as they appear in examination papers, are for the most part the merest fluff and ravelings of the adolescent mind, revealing neither the student's independent thought, nor, except casually, his command of English. They came into existence, I have been told, as a convenience for the examiner, who thus thought to combine in one paper questions on both sides of a pupil's training. The combination represents an accident of preparation, not an essential of secondary study [p. 14].

If one must have an entrance exam, says Scott, let it be sensible. Have a literary examination to see what the student has enjoyed and how profitable it has been to him. Use a composition examination that determines "how clearly [the student] can express himself on some subject in which he is undeniably interested and on which he is sure to have some information" (p. 17). Scott then delivers his harshest comment on the Harvard entrance examinations and the infamous reports of the 1890s, which they generated:

> I have recently been reading over again the reports of the Harvard Committee on English Composition, in which a number of examination papers are produced in facsimile, and the distorted English of the writers is almost indecently exposed. Upon this same English there is much sarcastic comment in the committee's report, and the exhibits seem to justify it; but for my part I could not view these reelings and writhings of the adolescent mind without a feeling of pity. It was all so unreal. Back of this mess and confusion were genuine individuals with likes and dislikes, with budding ambitions, with tingling senses, with impulses toward right and wrong. Where did these individuals come in when judgement was passed upon their faulty English? What were they trying to do? What motives lay behind these queer antics of the pen? If one could only tear away the swathings, set the imprisoned spirits free, and interrogate them, a strange new light might be thrown upon the causes of bad English.
>
> Another thought occurred to me as I read the reports. Should we not— at least those of us who are pragmatic philosophers—apply to the young offenders the crucial test of pragmatism? Where are they now, the writers of these rejected addresses? Are they in jail? Are they social outcasts? Are they editing yellow journals, or in other ways defiling the well of English? Or are they eloquent preachers, successful lawyers, persuasive insurance agents, leaders of society? I do not wish to pursue inquiries which may overturn the pedagogical foundations, but I am frankly curious to learn how far the actual course of events will bear out judgements based upon such evidence . . . the entrance requirements should throw the emphasis upon the things which are of most importance. It is of course necessary that our young people should spell and punctuate properly, should make the verb agree with its subject, should use words in their dictionary senses and write sentences that can be read aloud without causing unnecessary pain to the mandibles. They should also know the meanings of the words in the poetry and prose that they read, and understand the allusions to things ancient and modern. But these matters, after all, are subsidiary and must be treated as such. They are means to an end. To treat them as an end in and for themselves is to turn education in this subject upside down. The main purpose of training in composition is free speech, direct and sincere communion with our fellows, that swift and untrammeled exchange of opinion, feeling, and experience, which is the working instrument of the social instinct and the motive power of civilization. The teacher of composition who does not somehow make his pupils realize this and feel that all of the verbal machinery is but for the purpose of fulfilling this great end, is false to his trust [pp. 17–19].

It is difficult, within the limits of a chapter, to set down in detail Scott's particular contributions to rhetorical theory, but a few citations will suffice to justify Gertrude Buck's assertion that his greatest originality lay in this area. Three large purposes inform Scott's work in rhetorical theory: (1) his attempt to ground composition theory and practice in the rhetorical tradition from which it had become separated and to determine what in that tradition was still relevant to the modern world; (2) his attempt to enrich rhetorical theory with insights from other disciplines, particularly linguistics and psychology; (3) his interest in enlarging the nature and scope of inquiries into the uses of language.[16]

Scott knew the ancient tradition of rhetoric well, and it appears that his own personal integrity made him far more sympathetic to the ethical issues raised by Plato than to the exposition of rhetoric as a science by Aristotle. This emphasis comes through strongly in "Two Ideals of Composition Teaching," a paper he delivered before the Indiana Association of Teachers of English, 11 November 1911.[17] In essence, he said that "our teaching of composition and our attitude toward composition are still controlled, whether we know it or not, by the ingenious thought of the Sicilian Korax, as elaborated by the powerful mind of Aristotle" (SOAS, p. 38). And the aim of this kind of rhetoric "was success, and only that. Whether the speaker's cause was right or wrong, whether justice was being promoted or defeated, was no business of the teacher. Having put into the hands of the pleader the keen knife of persuasion, his task was ended. What the pupil did with his weapon after he got possession of it was a matter of no consequence to him" (SOAS, p. 36). Scott apparently could not accept the infrequent qualifications in Aristotle's work about the necessity of serving truth and justice and still saw there essentially an amoral art of persuasion being expounded in detail.[18]

Scott, of course, accepted Plato's representation of the Sophists as amoral rhetoricians, and he apparently saw Aristotelian rhetoric as sophistic in its morality. Modern scholars like James Kinneavy would not accept that position, and they are much more disposed to be tolerant to the ancient Sophists, feeling that Plato's representation of them was neither complete nor objective, and there is good reason for taking that position. However, when one considers the lack of integrity in modern advertising alone, it is not difficult to join Scott in sharing Plato's insistence upon the integrity of language.

Scott was also much taken with the principle of organic unity set forth in the *Phaedrus*, in which Socrates tells his young auditor that "any discourse ought to be constructed like a living creature, with its own body, as it were; it must not lack either head or feet; it must have a middle and extremities so composed as to suit each other and the whole work."[19] Scott made this conception the basis of his entire approach to the teaching of writing, the fundamental rationale for his argument that organic conceptions of structure were superior to mechanical. He called it "the profoundest idea of modern formal aesthetic—that of organic unity."[20] One may well wonder exactly what Scott meant by writing governed by mechanical rather than organic conceptions. The distinction is not difficult to understand. Most of the writing modern freshmen are encouraged to do is mechanically structured; the five-paragraph essay; the paper split into introduction, body, and conclusion, or beginning, middle, and end; the thesis statement and elaboration. The sections of such essays are indicated boldly; the joints between them are sharply etched. Organic structures are more subtle. They may be elaborate inductions such as Thomas Huxley's brilliant essay "On A Piece of Chalk," which begins with the chalk in an English carpenter's pocket and ends by affirming the validity of Darwin's theory of evolution. They may be discursive, such as the essays of Charles Lamb. To describe them, Scott drew on the analogy of the living plant, which grows and differentiates. The parts are apparent, but they are inseparable from the whole. An essay that grows and develops like a living plant would be organically structured.

Scott is also interesting to contemporary composition scholars because of his attempts to enrich rhetorical theory with insights from other disciplines, particularly linguistics and psychology. One measure of the extent to which he was doing this can be found in the work of his graduate students at Michigan, some of whose M.A. and Ph.D. theses appeared in a series of nine monographs Scott edited and titled "Contributions to Rhetorical Theory." They appeared over a period of twenty-three years, beginning with Gertrude Buck's M.A. thesis, "Figures of Rhetoric," in 1895 and concluding with Alice Snyder's "The Critical Principle of the Reconciliation of Opposites as Employed by Coleridge." "The Figures of Rhetoric" is an attempt to establish a sound psychological basis for the human tendency to employ figurative language. It is informed not only by great works in the tradition of rhetoric (Quintilian's *Institutes of Oratory*, Blair's *Lectures on Rhetoric and Belles Lettres*, and Whately's *Elements of Rhetoric* are cited) but also by James's *Principles of Psychology*, Höffding's *Outline of Psychology*, Dewey's *Psychology*, Bain's *Senses and Intellect*, Wundt's *Human and Animal Psychology*, Ladd's *Psychology, Descriptive and Explanatory*, and Baldwin's *Handbook of Psychology, Senses and Intellect*.

Buck's Ph.D. dissertation, "The Metaphor—A Study in the Psychology of Rhetoric," the fifth of these monographs, is discussed at length by Albert Kitzhaber. He notes that "it was distinguished sharply from all previous discussions of the subject by an attempt to use the data of experimental psychology in shedding light on the origin, nature, and use of this figure" (p. 282). He concluded by observing that Buck

deserves much credit for her courage in brushing away the accumulated
dust of well over two thousand years and for making an earnest attempt to
attack the subject of metaphor from a wholly new point of view, bringing
to bear on it the most recent and most pertinent information she could
find, even though that knowledge was drawn from another discipline. In a
period when rhetorical theory was becoming steadily more isolated from
other fields of knowledge, Buck's attitude was unusually and commenda-
bly independent [p. 291].

In 1922, in "English Composition as a Mode of Behavior," Scott
himself brought to bear on composition teaching some early principles
of behavioral psychology. Too much schooling proceeds, he said, as if
the child who enters school were

> a great emptiness to be filled and a great dumbness to be made vocal.
> Ignoring, or at least undervaluing, the gestures and poses and cries and
> modulations that are the child's natural medium of expression, the
> teacher proceeds to unload upon him the colossal structure of our
> speech—one of the most complicated, the most ingenious, the most
> abstract, and most delicate of all of the creations of human reason. In
> almost all respects it is at the opposite pole from the language that he is
> accustomed to in practice. . . . If it is dissociated from the child's instinc-
> tive language, if new habits of speech and speech-control are set up in
> this higher plane as if there were no habits in the lower plane already in
> possession, then it is being used injudiciously and trouble may be ex-
> pected to follow [SOAS, pp. 26, 27].

Scott described the conflict students experience when deep-seated
language habits conflict with those attempting to be imposed from
above as "disastrous collisions and derailments" (SOAS, p. 27).

> A child at such a stage often goes through a series of somnambulistic
> convulsions, as if he were trying to throw off a leaden weight. His whole
> scholastic life is an effort to escape from what seems to him the body of
> death. Vacation is welcomed as a happy release, and the thought of being
> put on the rack again makes him miserable [SOAS, p. 27].

Many of the problems we see in English composition proceed from
this conflict between a student's natural language habits and those he
or she is attempting to employ to please the teacher.

> The maddening errors that students are guilty of—the verbless sentences,
> the ludicrous malapropisms, as for example, "The Chinese are Confu-
> sionists in religion"—the unconscious lapses into slang, the sudden
> plunge from the sublime to the ridiculous—these are not, in any due
> proportion, the product of stupidity, or malice, or even laziness. They
> are the outward signs of an inward lesion—of the disjunction of two
> phases of man's nature that can work as they should only when they
> work together [SOAS, p. 28].[21]

As one would expect, Scott was thoroughly familiar with the works of contemporary linguists, and he was solidly on the side of those who, like Jespersen, for example, argued for a descriptivist position. It is no accident that twentieth century leaders in the movement for more liberal and scholarly attitudes toward usage—one thinks immediately of Sterling Leonard, Ruth Weeks, and Charles Fries—were graduate students in Scott's rhetoric program at Michigan early in the century.

Scott's interest in extending the range of topics that students of language and rhetoric usually consider can be indicated by brief summaries of a few more of his papers. For example, "The Genesis of Speech," his presidential address at the 1907 MLA Convention, was an attempt to establish a link between normal physiological processes and the earliest meaningful human utterances. To develop his argument, Scott dipped extensively into the work of language historians and physiologists. His literary papers reveal an interest in vowel alliteration in modern poetry, a subject he felt had been neglected; an attempt to discover the basic differentiating features of poetry and prose; and his interest in the complex and elusive patterns of prose rhythms.

When one scans, even cursorily, Scott's long career and the range of topics that interested him, it becomes apparent quickly that here was the Renaissance man of his era. He brought to the field of English the intellectual equipment, knowledge, and curiosity that have always distinguished first-rate minds, but his serious scholarly interest in rhetoric and composition was shared by so few of his contemporaries that he was forgotten soon after his death. Louis Strauss, head of Michigan's English Department and a colleague of Scott for thirty years, summed him up better than anyone else has yet been able to do. Justifying the merger of Michigan's Rhetoric and English Departments in 1930, Strauss reviewed the events that led to the formation of the Department of Rhetoric in 1903:

> . . . in Professor F. N. Scott, the University possessed an educational leader of national, indeed of world reputation. . . . Ten years earlier he had burned his boats and declared himself wholly and exclusively committed to rhetoric as his field of investigation and instruction—a daring venture at that time, when it was anything but a popular subject, and utterly without standing as a field of graduate study and research. Professor Scott speedily gave it such standing.

Students who came to Scott's rhetoric program were never disappointed, continued Strauss:

> Professor Scott was the ideal teacher of the science and art of literary expression. His range of interests was well-nigh universal. An acute critic, an accomplished linguist, a skillful bibliographer, he had at command, beside his splendid background in English literature and his thorough

mastery of the history of rhetoric, a wide knowledge of the arts of painting
and music and of many literatures, ancient and modern, a profound grasp
of philosophy in general and aesthetics in particular, and such an appreciation of many sciences as few teachers of the humanities possess. . . . Dr.
Scott's conception of rhetoric was catholic in the extreme; it was limited
only by the range of his own personal interests, which really means that it
was not limited at all. Under the spell of his magnetic and stimulating
personality his students developed to their utmost capacity. They are to be
found everywhere—brilliant teachers, successful writers, and men in every
walk of life upon whose tastes and characters his influence is indelibly
stamped; and they are not backward in saying so.[22]

Strange that our profession lost track of a man of such distinction for
so many decades.

Notes

1. Letter in the Fred Newton Scott Papers, Michigan Historical Collections,
   Bentley Historical Library, Ann Arbor, Michigan. (This collection has approximately 7,500 items, primarily correspondence. It has recently been
   augmented by a file in my name. The latter contains taped interviews with
   two of Scott's living children and several of his former students.) Buck had
   taken her M.A. in 1895, her Ph.D. in 1898, and then gone on to a successful teaching career at Vassar. Miss Jackson, who had achieved a successful
   career in magazine editing and writing, gave that up when she became
   Scott's personal secretary and then second wife (in 1924) after the death of
   the first Mrs. Scott in 1922. She proved to be a faithful and caring com-
   panion to him during the last seven years of his life, when failing health
   and finally arteriosclerosis of the brain reduced him to helpless invalidism.
      The occasion of the letter was a memorial to Scott that Georgia was
   trying to define and get underway. The project reached fruition many
   years later as *The Fred Newton Scott Anniversary Papers,* a collection of essays
   by Scott's former students and colleagues, edited by Charles Whitmore and
   Clarence Thorpe, and presented to him at Ann Arbor's Barton Hills
   Country Club in July 1929—less than two years before his death.
2. This information appears in an obituary Scott wrote at the death of his
   sister Harriet in 1904. In it he incorrectly referred to Mr. Jones as Walter
   A., but his gratitude to the man was unequivocal and sincere: "He [Jones]
   even had the temerity to put the new science of psychology into the training school, for which I bless him forever" ("Biography of Miss Harriet M.
   Scott" in the Fred Newton Scott Papers).
3. I wish to acknowledge here the considerable and very kind assistance given
   to me in gathering materials about Scott's personal life by his only
   daughter, Marion Scott Goodrich; his youngest son, Richard; and his
   granddaughter, Marilyn Guenther, the daughter of Mrs. Goodrich. I
   should note further that I have resisted supplementing this account of
   Scott's early years with the information contained in John Dewey's sketch

of him because of inconsistencies in the latter that I cannot fully resolve at this time. For example, Dewey, who must have got his information from Scott (his younger colleague at Michigan in the early 1890s), says that Scott received his education in the public schools of Terre Haute. But he also reports Scott's early interest in science and then his growing interest in languages and classical studies stimulated by his German teacher at the Normal School. It is not clear to me whether Dewey and Scott used "public" and "normal" synonymously or whether they were different institutions. And I cannot account for the fact that Scott apparently did not mention Jones to Dewey, especially because of Dewey's involvement in that subject. See John Dewey, "Fred Newton Scott," *The Early Works, 1882–1898*, vol. 4, ed. Jo Ann Boydston (Carbondale: Southern Illinois University Press, 1971), pp. 119–23.

4. W. R. Humphreys, "The Department of English Language and Literature," in *The University of Michigan, an Encyclopedic Survey in 9 Parts*, Part II (Ann Arbor: The College of Literature, Science, and the Arts, 1943), p. 550.

5. For Demmon's actual note and the resolution it prompted, see Michigan University Board of Regents Proceedings, April 1903, Michigan Historical Collections, Bentley Historical Library, Ann Arbor, Michigan.

6. Among the groups to which he belonged were the Quadrangle Club and the Azazels on the Michigan campus; the Linguistic Society of America, of which he was a founding member; and the English Association of Great Britain; the American and British Association for the Advancement of Science; and the Philosophical Society of Great Britain.

7. Letter from Ray Stannard Baker to Willard Thorpe, 20 December 1928, in the the the Fred Newton Scott Papers.

8. See Donald C. Stewart, "The Barnyard Goose, History, and Fred Newton Scott," *The English Journal*, 67:8 (November 1978), pp. 14–17; "Rediscovering Fred Newton Scott," *College English*, 40:5 (January 1979), pp. 539–547; James Berlin, *Writing Instruction in Nineteenth Century American Colleges* (Carbondale: Southern Illinois University Press, 1984), pp. 77–84.

9. Preface, 3d ed. (Boston: Allyn Bacon, 1895), p. iii.

10. See Alexander Bain, *English Composition and Rhetoric* (New York: D. Appleton and Co., 1866), Pt. I, chap. 5; A. S. Hill, *The Foundations of Rhetoric* (New York: Harper and Brothers, 1892), pp. 305–325, and *The Principles of Rhetoric* (New York: American Book Co., 1878), pp. 157–161; D. J. Hill, *The Elements of Rhetoric and Composition* (New York: Sheldon and Co., 1878), pp. 71–77, and *The Science of Rhetoric* (New York: Sheldon and Co., 1877), pp. 198–202; J. S. Clark, *Practical Rhetoric* (New York: Henry Holt and Co., 1886), pp. 28–32; T. W. Hunt, *Principles of Written Discourse* (New York: A. C. Armstrong and Son, 1884), pp. 82–84; G. R. Carpenter, *Exercises in Rhetoric* (New York: Macmillan, 1893), chap. 12; J. G. R. McElroy, *The Structure of English Prose* (New York: A. C. Armstrong and Son, 1885), pp. 196–222; John F. Genung, *The Practical Elements of Rhetoric* (Boston: Ginn and Co., 1886), pp. 193–213; and Barrett Wendell, *English Composition* (New York: Charles Scribner's Sons, 1891). Scott

and Denney call Genung's "an exhaustive analysis, the best that has yet appeared" (p. 106).

11. For full citations of all works by Scott cited in this essay, see the list of selected publications accompanying this paper. Quotations from them will be indicated by page numbers in parentheses.

12. For more recent discussions of these topics, see Albert R. Kitzhaber, "Rhetoric in American Colleges: 1850–1900," dissertation, University of Washington, 1953, pp. 191–223; Robert Connors, "The Rise and Fall of the Modes of Discourse," *College Composition and Communication*, 32: 4 (December 1981), pp. 444–464.

13. Scott and Denney do not give a full citation for this work.

14. In a footnote to this remark, they cite the Preface to Bosanquet's *Logic*, Vol, I, p. vii, for a similar conception of judgment-forms. They say, also, that Bosanquet acknowledges indebtedness for the idea to Mr. Alfred Robinson, of New College, Oxford, p. iv. The metaphor of the plant here must be especially startling to those familiar with Gordon Rohman and Albert Wlecke's *Pre-Writing: The Construction and Application of Models for Concept Formation in Writing*, Cooperative Research Project No. 2174 (East Lansing: Michigan State University, 1964), pp. 11–14. Criticism of this conception in Scott's work occurs in Allan Seager, *The Glass House: The Life of Theodore Roethke* (New York: McGraw, 1968), p. 51. For a rebuttal of Seager and of H. L. Mencken, who made numerous disparaging but irresponsible and inaccurate remarks about Scott and his work, see Donald C. Stewart, "Reputation Lost: A Brief Note in the History of American Letters," *Menckeniana*, No. 85 (Spring 1983).

15. See particularly, Kitzhaber, op. cit., pp. 71–79; A. S. Hill and Elizabeth A. Withey, "Sub-Freshman English," *Educational Review*, 14 (June–December 1897), pp. 468–495; 15 (January 1898), pp. 55–73.

16. For additional discussion of Scott's originality in these matters, see Kitzhaber, op. cit., 113–120; Berlin, op. cit., pp. 101–112.

17. It was published in the *Proceedings* of that association for that year. Passages quoted here are from the reprint of the article in *The Standard of American Speech and Other Papers* (1926). Page numbers of subsequent quotations from articles in this book will be preceded by *SOAS*.

18. On 12 April 1926, Charles Sears Baldwin wrote to Scott, telling him, "I have been sitting by, all last evening with your essays (the essays in *The Standard of American Speech*), bridling at your unretracted distrust of Aristotle, chuckling, admiring, learning." The tone of the entire letter is warm and congenial, but it is interesting to note that Baldwin knew Scott's estimate of Aristotle and probably wished he could be persuaded to change it (letter in the Fred Newton Scott Papers).

19. Plato, *Phaedrus*, trans. R. Hackforth (Cambridge: Cambridge University Press, 1972), p. 128.

20. For a fuller exposition of Scott's thinking on this subject as well as his indebtedness to Plato, see "Rhetoric Rediviva," ed. Donald C. Stewart, *College Composition and Communication*, 31:4 (December 1980), pp. 413–419.

21. No one of our contemporaries, I believe, has developed the implications of

what Scott is saying here as fully as Robert Zoellner. It was he who alerted us to students' "tragic proficiency in writing themes made up of *words-for-teacher* which are seldom if ever *words-for-me*," a dissociation, says Zoellner, of what the student actually thinks and even says when he is speaking comfortably with his peers and the language he or she uses in the classroom setting. Ironically, when Zoellner identified this problem forty-seven years *after* Scott, his monograph was regarded with intense hostility or incredulity. See "Talk-Write: A Behavioral Pedagogy for Composition," *College English*, 30:4 (January 1969), pp. 305–307.

22. Louis A. Strauss, "Regents Merge Two Departments," *The Michigan Alumnus*, 36 (1929–30), pp. 331–332.

Selected Publications

Scott, Fred Newton. "College Entrance Requirements in English." *School Review*, 9 (June 1901), pp. 365–378.

Scott's second comment on entrance requirements in two years, this one laying out differences in what he calls feudal versus organic conceptions of the nature of a university and what it should require of those who enter it.

———. "The Report on College Entrance Requirements in English," *Educational Review*, 20 (October 1900), pp. 289–294.

Scott's statement of dissatisfaction with a report on college entrance requirements that fails to establish a number of relationships between the nature and purpose of entrance requirements and the aims and methods of English teaching at that time.

———. *The Standard of American Speech and Other Papers*. Boston: Allyn and Bacon, 1926.

A collection of Scott's previously published papers on a wide range of topics. Of most interest to composition and rhetoric specialists are the following: "The Accentual Structure of Isolable English Phrases," "English Composition as a Mode of Behavior," "The Genesis of Speech," "The Most Fundamental Differentia of Poetry and Prose," "The Order of Words in Certain Rhythm-Groups," "Our Problems," "The Scansion of Prose Rhythm," "The Standard of American Speech," "Two Ideals of Composition Teaching," and "Vowel Alliteration in Modern Poetry."

———. "What the West Wants in Preparatory English." *School Review*, 17 (January 1909), pp. 10–20.

Scott's harshest comment on entrance requirements in eastern universities and on the Harvard Reports of the 1890s.

———(ed.). *Contributions to Rhetorical Theory*. Published from 1895 through 1918.

Series of monographs written by Scott's students and colleagues (with one exception), and edited by Scott. Most were published in Ann Arbor by the Ann Arbor Press. Scott's contribution, the fourth of these monographs, was an extensive bibliography of practical and theoretical works entitled "References on the Teaching of Rhetoric and Composition." No date is given, but as Scott is listed as a junior professor at the time of its publication, we know that it must have been between 1896 and 1901. Gertrude Buck's "Figures of Rhetoric: A Psychological Study" (1895) and "The Metaphor—A Study in the Psychology of Rhetoric" (1899) were the first and fifth monographs in this series.

Scott, Fred Newton, and Gertrude Buck. *A Brief English Grammar*. Chicago: Scott, Foresman, 1905.

A textbook for English grammar written in collaboration with his first Ph.D. candidate in rhetoric at Michigan. The book features an organic conception of the sentence and takes an essentially descriptive position on the question of correct usage.

————, George R. Carpenter, and Franklin Baker. *The Teaching of English in the Elementary and the Secondary School*. New York: Longmans, Green, and Co., 1908.

Scott's best statement on the necessary qualifications for one wishing to enter the English teaching profession. Also contains stimulating discussion of ways to enliven composition and literature instruction.

————, and J. V. Denney. *Aphorisms for Teachers of English Composition, and the Class Hour in English Composition*. Boston: Allyn and Bacon, 1905.

A collection, somewhat random, of wise observations about the way writing should be taught.

————, and ————. *Composition-Literature*. Boston: Allyn and Bacon, 1902.

A traditionally organized text, which stresses the need for would-be writers to study the work of those who do it well, to write on subjects in which they are deeply interested and to which they are strongly committed, and to attempt to develop a natural (or authentic) voice. Revised and published under the title *The New Composition-Rhetoric* in 1911.

————, and ————. *Composition-Rhetoric, Designed for Use in Secondary Schools*. Boston: Allyn and Bacon, 1897.

A text stressing the idea of composition as living, organic, kinetic, not static. Emphasis also on synthesis more than analysis. Introduces some early brainstorming techniques.

————, and ————. *Elementary English Composition*. Boston: Allyn and Bacon, 1900.

Still another collaboration between Scott and Denney, this one emphasizing the social context in which the student writes and urging heightened awareness of that fact as a means to improving student motivation for writing.

————, and ————. *Paragraph-Writing*. Boston: Allyn and Bacon, 1893.

Essentially the second edition, much enlarged, of a pamphlet originally prepared by the authors in 1891. It features the concept of the paragraph as a theme in miniature, presents a theory of the paragraph in terms of current psychological theory, and draws on a number of conceptions in paragraph theory at that time. Probably Scott's most successful and continuously used text.

————, and Gordon Southworth. *Lessons in English,* Books I and II. Boston: Benjamin H. Sanborn & Co., 1906.

An arranged collaboration between Scott, a university man, and Southworth, an experienced public school teacher and administrator. Essentially traditional texts on grammar and composition that break no new ground.

I. A. RICHARDS

by Ann E. Berthoff

When I. A. Richards died in the fall of 1979 at the age of eighty-six, the encomiums did not include substantial appreciation of any books written in the last twenty-five years of his life. Of course, considerations of space had something to do with it. Giving even a bare account of that extraordinary career necessitated citing his work in the theory and practice of literary criticism, his efforts to ensure acceptance of Basic English as a model for language instruction, and his contributions to thinking about ways and means to facilitate the international exchange of ideas on which the survival of civilization depends. What room there was left went to an acknowledgment of the fact that he had taken up mountaineering in his youth and the writing of poetry when he was an old man.

But if the obituaries were disappointingly thin, they only echoed the curious kind of recognition Richards had had during a lifetime as a scholar working in several fields. Stanley Edgar Hyman, in his review of the accomplishments of modern literary criticism (*The Armed Vision* [1948]), expressed sadness (and a little contempt) for the direction Richards had begun to take away from textual criticism and rhetorical theory toward pedagogy. But there was no rejoicing when he got over into educationland. Recognized as an important figure "generally speaking," Richards was never welcomed vigorously by any one discipline claiming him as its own; indeed, it was sometimes the case that he was completely ignored. Linguists paid no attention whatsoever to his passionate but reasoned criticism of all they were about. Seymour Chatman observed in a 1957 review of *Speculative Instruments* in *Language* that it was the first time that that journal, the official organ of the

Linguistic Society of America, had noticed a book by I. A. Richards. (And the society never let it happen again.) Richards wrote continually about matters of the most urgent concern to the teaching profession, but the teaching profession in its journals and conferences and conventions acted as if he did not exist. Nowadays it is hard to find a graduate student or a young instructor who has any notion of the role Richards played in shaping modern criticism, to say nothing of his aims in education. A professor at Princeton has told me that in a seminar on literary theory his students are fascinated by Lacan and Derrida but have never heard of *The Meaning of Meaning* or *The Philosophy of Rhetoric*.

Nevertheless, it is certainly true that Richards has had more influence than any other critic in formulating the principles of close reading that the New Critics took as their method and, perhaps what is more important, in providing a rationale for the development of such a theory of reading in the first place. *Practical Criticism* (1929), with its demonstration of the actual work of seemingly competent readers, shocked everybody into the recognition that it was foolhardy to assume that the primary aim of English studies was to improve taste, the ability to read for sense and meaning being a foregone conclusion. A generation of teachers—and the teachers they taught—learned from *Practical Criticism* and *The Philosophy of Rhetoric* how to teach close reading.

There have been as yet no comparable results from the second half of Richards' career, beginning with the time he moved from Cambridge, England, to Cambridge, Massachusetts; but there is a good chance that his influence will one day be felt, simply because he understood profoundly the importance of literacy training and developed procedures of demonstrable effectiveness. Richards' pedagogy evolved in the widest possible context—world history. He saw today's illiterates standing before the modern world in much the same way as the Homeric Greeks had stood on entering historical times. His inventiveness kept pace with his sense of urgency, and he never wavered in his belief that reading and writing are moral concerns and that what we do in teaching them has a vital political dimension. Learning to choose words and deciding the way we want them to work can become "an introduction to the theory of all choices."[1] Gains in literacy are, therefore, gains in human freedom. Closer to home, Richards' contributions to reading theory, English as a Second Language (ESL) pedagogy, and an appreciation of the opportunities and hazards of electronic media are too sound to be permanently disregarded. His education design is so full of good sense, is so well founded philosophically and linguistically that it is bound to survive despite the dismissal of those "techniques of language control" by which it was meant to be put into practice.

Whatever influence Richards has had on the teaching of writing has been oblique, coming by way chiefly of his theory of critical reading.

Teachers who have themselves learned to read according to the precepts of the New Criticism are likely to have the habit of reflecting on the reflections language makes possible and thus to be able to teach composition by showing their students how to look and look again, how to explore how words work, how to think about thinking. This is not to say that old-fashioned teachers of literature are automatically good writing teachers—only that a sound method of critical reading is adaptable to the teaching of writing. Indeed, the old formula—students can only write as well as they can read—may be reversed as teachers come to understand what it means to say that writing is a mode of learning: *students can read only as well as they can write.*[2]

My claim in what follows is that there is a composition pedagogy of great power to be developed from books that have been forgotten or ignored for forty years by the confraternity of rhetoricians and professional educators. I believe that for those who want to teach writing as a means of making meaning and to teach others to do so, it will be illuminating to survey Richards' entire career. His aim was from the first to find ways to give an account *of* meaning and to account *for* meanings; to move, as the title of one of his important essays has it, "towards a theory of comprehending." Surely that is what we most urgently need if we are to teach writing as a mode of learning and a way of knowing.

I cannot, of course, fully describe here Richards' massive output—over twenty-five volumes—but I hope to show how each of the major books grew from those preceding it, as Richards sought ways to bring his own theory to bear on practice in the classroom, which he liked to call a "philosophic laboratory." In order to explain how his developing philosophy of language steadfastly informs his conception of rhetoric, I have grouped as follows those books of Richards to be discussed according to three stages of his career, roughly identified by the places where he was teaching.

Cambridge University (1923–1929): *The Meaning of Meaning, Principles of Literary Criticism, Science and Poetry, Practical Criticism.*

China (intermittently, 1929–1938): *Mencius on the Mind, Coleridge on Imagination, The Philosophy of Rhetoric, Interpretation in Teaching.*

Harvard University and Cambridge, Massachusetts (1938–1974): *How to Read a Page, Speculative Instruments, So Much Nearer, Design for Escape, Beyond.*

I. A. Richards came to teaching only after having set aside mountaineering and psychoanalysis as possible careers. It's a fair claim that he conceived of pedagogy from the first as presenting the same kind of intellectual challenge as the foregone professions offered, requiring careful diagnosis and bold reconnoitering. If the hazards and risks were comparable, so was the exhilaration of seeing things through. Teaching was at first ancillary to theorizing; it provided the grist. And if later the roles of theory and practice were reversed, the important point is that Richards never lost sight of either one. The shift in the middle of his life from literary criticism to educational design—"from criticism to creation," as he put it—is not so puzzling as it might appear if it is remembered that Richards undertook no study without a practical—that is to say, pedagogical—purpose in mind.

Richards took the first of his three Cambridge degrees in 1915 (the B.A. was followed by the M.A. and Litt. D) and from 1922, when he was appointed Lecturer in English and Moral Sciences, until 1929, when he left for the Orient, he was associated with Magdalene College. William Empson, his most famous student, has described his lecturing style:

> More people would at times come to his lectures than the hall would hold, and he would then lecture in the street outside; somebody said that this had not happened since the Middle Ages, and at any rate he was regarded as a man with a message. There were those who called him a spellbinder, implying that there must be something wrong with a lecture if it produced this effect. . . . [3]

Following the publication of several papers and pamphlets on aesthetics and symbolism, Richards' first book appeared in 1923—*The Meaning of Meaning,* which he wrote with C. K. Ogden. It is a young man's book—assured, brash, hasty, careless, brilliant. (Richards himself was of this opinion: "When [Ogden] and I were, in the exuberance of youthful ambition, throwing together *The Meaning of Meaning. . . .*") Certain of its explanations and demonstrations have long since been shown to be inadequate and misleading, but the book deserves its reputation as a herald, and its appearance was certainly a sign of the times. Cambridge, in the years directly after the First World War, had reformed English studies, with consequences that reverberate still.[4] Anglo-Saxon and Middle English were dropped as degree requirements; contemporary literature and up-to-date critical and philosophical issues supplanted purely philological concerns. In this atmosphere, a discussion of the meaning of "meaning" was bound to find an audience; Richards certainly knew how to engage their interest. Here is another glimpse of the spellbinder:

> At first he was not technically a good lecturer. He turned his back on the audience too frequently and for too long, while he scribbled on the board scarcely legible phrases, or scarcely intelligible diagrams. He broke many pieces of chalk . . . [But] I. A. R.'s lectures were spell-binding, partly because we could not fail to notice that he was breaking not only chalk but new ground.[5]

The new ground broken in *The Meaning of Meaning* has by now been so thoroughly explored and developed that its original outlines are not readily discernible, but in fact Ogden and Richards were among the first to raise questions about context and interpretation, to identify the variety of language functions, to differentiate *reference* and *referent*— and to investigate the meanings of *meaning*. The authors declare that their interest is in developing a *science* of symbolism, an endeavor for which grammarians, philologists, metaphysicians, and most philosophers offer no guidance. Some who do are represented in appendixes that provide basic texts from Husserl and Frege on *Sinn* and *Bedeutung*, Russell on the logic of denotation, and Peirce on the triadic nature of the sign. Included as a supplement is an essay by Malinowski, "The Problem of Meaning in Primitive Languages." This account of how the "context of situation" must be understood if an investigator is to interpret the meanings intended by Trobriand Islanders as they go about their fishing and competitive sailing provided "scientific" support for Ogden's and Richards' fundamental concept of the *sign-situation*. Malinowski, in turn, praises the authors of *The Meaning of Meaning* for their insight into the kinds of questions that should guide the field study necessary to the new science of symbolism. His essay provides an early example of the cross-fertilization of linguistics and ethnography that characterizes modern anthropology.

In years to come, it was Richards, not Ogden, who saw the implications of an emphasis on interpretation, on comprehending and discovering further purposes in the process of making meaning. As Richards moved away from positivist notions of what constitutes semiotics, from narrow ideas of what the "science" of symbolism entails, towards conceptions of mind that in Ogden's view were "metaphysical," a break was inevitable. Richards remarked on the irony of the fact that two people who at first understood one another completely within twenty years "couldn't write a line to one another without grossly misunderstanding."[6] Nevertheless, Ogden's influence on Richards was in many ways long-lasting and beneficial. From his early collaborator, Richards learned (and never forgot) the importance of the concept of opposition[7] and the related conception of the role of diagrams and all kinds of schematic representation. He saw how valuable it was to work always from and towards concrete examples, the actualities of language use. Most important, he found in Ogden's Basic English the model of the auxiliary instruments needed in teaching reading and writing, whether to native speakers or to those learning a second language.[8] And from the very first, the political dimensions of literacy were explicitly recognized and delineated.

By the time *The Meaning of Meaning* went into a second edition three years later, Richards had written *Principles of Literary Criticism* (1924) and *Science and Poetry* (1925), in which he tried to do for the *emotive* function of language what the earlier book had done for the *symbolic*— to provide "a critical foundation." This attempt to formulate a science of poetry led him into swamps from which he struggled mightily to extricate himself ever after. (Fifty years later, he was still trying to explain what he had meant by "pseudo-statement.")[9] But Richards never repudiated these early books, though he did publish an extensively annotated edition of *Science and Poetry,* retitled *Poetries and Sciences* (1970). Of *Principles of Literary Criticism,* he wrote:

> I regard it still with a benevolent eye as being a better sermon than it knew itself to be. And its attempts to put the causes, characters, and consequences of a mental event in the place of what is known, felt, or willed was *as science,* I think still, on the right lines. It seemed to many readers unintelligible then. Today to similar readers it might seem obvious [*Speculative Instruments,* p. 44n].

This characterization usefully reminds us that Richards' aim was to supplant faculty psychology with a version of pragmatism; he attempted to adapt the ideas he'd gathered from William James's *Principles of Psychology,* to reintroduce them as principles of literary criticism. In this first attempt to found the science of symbolism on a psychology of language, Richards thought that by clearing out the blurry old terms like *knowing* and replacing them with crisp new language like *causation,* he could demystify the relationship of "an awareness and what it is aware of." Epistemology seemed mostly irrelevant to the principles of literary criticism: "A theory of knowledge is needed at only one point . . . at which we wish to decide whether a poem is true . . . or reveals reality . . . , whereas a theory of feeling, of emotion, of attitudes and desires . . . is required at all points of our analysis" (*Principles,* p. 91).

Criticism was considered to have two aspects, communication and evaluation; psychology could take care of both. It shows how our proper experience of the poem is a matter of getting "the mental conditions relevant to the poem"; they must be consonant with "the mental operations" of the poet (*Practical Criticism,* pp. 10, 7). But if we have read the poem aright, there will be no need to run through a checklist of criteria in order to judge either our experience or the poem. "Getting the experience" makes evaluation supererogatory as a separate process. Psychology was important because it could help us understand that responses were neural events, actual operations that we could learn to carry out efficiently, "correctly." Richards depicts the perception of poetic diction in a schema aswarm with coils, coordinates, wavy lines, and directional arrows; metaphysical dilemmas disappear because graphic linkages span the gap between object and subject, that same gap that opens up between stimulus (*S*) and response (*R*) and that behaviorists fill in with little *s*'s and *r*'s, as Jerome Bruner once put it.

Richards was to discover very quickly that his principles did not go deep enough; it was classroom experience that brought him to this insight. In what is probably the first genuine research in the teaching of English, Richards gave his students unidentified poems—"texts" they would now be called—to which they were to respond anonymously and at their leisure over the period of a week. He then lectured the following week, the responses serving as his points of departure. This experiment, as he called it, provided the material for a "natural history of human opinions and feelings." He published his findings and his interpretations of their significance in *Practical Criticism,* a further contribution to that science of symbolism for which *The Meaning of Meaning* had called. Richards obviously delighted in the fact that the "scientific" evidence was derived from the study of poetry. He organized the responses, which he called "protocols," according to the attitudes exemplified. Each sort was decimalized, for example, 1.11, the first number referring to the poetic specimen; the second, to the attitude; the third, to the sample response. This was a sly allusion to the itemized propositions of the *Tractatus Logico-Philosophicus,* a way of dramatizing Richards' claim that what were needed were specimens, samples of actual responses, not mere theoretical assertion.[10]

Surveying the stunning exhibition of misreading gathered in *Practical Criticism,* Richards declared that what was certainly needed was a theory of interpretation; his principles of literary criticism, as I have noted, did not include such a concept. He faced squarely the evidence that among the "ten difficulties of criticism," none was more fundamental than the failure to construe, to read for sense. Richards explicitly noted that this and other deficiencies were not "inalterable defects of mind" and that they were remediable, that "method in techniques of reading" was clearly what was needed. It was absurd, he wrote, that these students, the best and the brightest, had not received such training. Richards speculated that with the disappearance of small communities in which the acquisition of language entailed command of its functions, there was a need to redress the bad effects of stereotypes encouraged by modern communication and to supplant the learning that used to go on outside the school. The need for a theory was thus premised on this socially and politically defined need to respond to the changing patterns of British national life.[11]

With guidance from a theory of interpretation, we would see that "the wild interpretations of others must not be regarded as the antics of incompetents, but as dangers that we ourselves only narrowly escape, if, indeed, we do. . . . The only proper attitude is to look upon a successful interpretation, a correct understanding, as a triumph against odds" (p.

315). Richards also meant that interpretation should itself be taught, adding characteristically, "Only the actual effort to teach such a subject can reveal how it may best be taught" (p. 317). Thus, pedagogy was seen as requiring the guidance of theory, which must, in turn, be examined in the light of what actually goes on as students read and write. Richards was the first teacher to treat student writing as a text deserving and repaying close attention; written responses and the careful study of those responses provide occasions for teacher and students alike to identify and evaluate ways and means of making meaning. *Interpretation in teaching* was to become for Richards a model whereby one could understand interpretation as the central act of mind.

"Practical criticism" is a phrase of Coleridge's; it serves to remind us that although the book that bears the title is generally paired with *Principles of Literary Criticism*, it also bridges to the next two studies, *Mencius on the Mind* (1932) and *Coleridge on Imagination* (1934), which were written during the decade Richards was intermittently in China. In the course of his activities as director of the Orthological Institute (Basic English) of China and as visiting professor at Tsing Hua University, Richards found in translation theory a challenge comparable to that posed by the Cambridge protocols. What happened between *Practical Criticism* and *Mencius on the Mind* was that Richards came to see that a theory of interpretation needs an adjunct theory of knowledge. Because the mental receptivity of readers cannot be monitored except by a study of how they read (or, more accurately, how they have *written* about what they have read), the next step would have to be a development of techniques to improve that reading, to try to account for misreading otherwise than by considering the functioning of the nervous system. Richards realized that his findings in *Practical Criticism* actually rendered his "scientific" principles of literary criticism obsolete; he came to see that criticism would have to concern not just the psychology of response but the laws of the mind. In *Mencius on the Mind*, problems of translation offer a focus for thinking about the laws of the mind as they are manifested in the movement from one language to another.[12]

Richards knew that just as the relationship of word and referent had to be mediated by *reference*, the idea being represented, so the relationship between what is said in one language and how that can be represented in another language entails interpretation of what is meant, and that, in turn, requires having alternate readings to compare. Thus, theory and practice are once again considered dialectically: "The very process of deciding between rival possible readings is in need of examination. Before this process can be studied it must be exhibited. And our purpose here should go beyond the mere elucidation of the passage in hand; it should aim . . . at a generalized technique by which meanings of all kinds on all necessary occasions can be systematically displayed" (*Mencius*, p. 28). Richards is very explicit about the necessity of bringing a theory of translation to the test: "New hypothesis is but old dogma writ large. Unless we do actually and constantly sketch out alternative definitions using different logical machinery we shall not gain the ability to experiment in interpretation which comparative studies require" (p. 90). I will return to the techniques first demonstrated in *Mencius on the Mind*, but I want here to sketch how a theory of knowledge developed from a theory of translation.

In attempting to account for Mencius' lack of interest in analysis or, more particularly, in analogy as a route toward logical analysis, and in considering such peculiarities of Chinese as the absence of tense, Richards had speculated on the possibility of minds actually and constitutionally being different. But philosophical differences redefined as socially determined attitudes that then affected somehow the constitution of mind were no more accessible than "mental operations." Richards never entertained these ideas further; it was his study of Coleridge that brought him to reaffirm, rather, the idea of the "*all in each* of human nature" (*Coleridge,* p. 97). Coleridge is a presence throughout *Principles,* but it was not until *Mencius,* where he had worked with passages that he thought "paralleled" Coleridge, that Richards found what the father of modern criticism had to teach him: how to think about thinking. In *Coleridge on Imagination,* Richards told Reuben Brower, he "wasn't so much concerned to say what Coleridge had thought as to suggest what might be done with what he had said."[13] What Richards did was to find a rationale for focusing not on mental operations (which were, in any case, inaccessible) but on texts as we come to interpret them, to know what their authors are saying.

The motive for shifting from the psychology of response to the interpretation of texts had been sounded in *Practical Criticism:* "Psychologists have never, I think, resolutely faced this question of how we know so much about ourselves that does not find any way at present into their textbooks. Put shortly, the answer seems to be that that knowledge is lying dormant in the dictionary. Language has become its repository, a record, a reflection, as it were, of human nature."[14] Thinking with the concept of imagination, Richards concluded that the laws of mind could be formulated by reflecting on the reflections provided by language. Thus, the gatherings and connections of sense organs, reality, remembered experience, and so forth, which he represented earlier in diagrams of the nervous system, are supplanted by the interdependencies of meanings, which constitute the active mind. In recognition that that is a philosophical matter, Richards called the book in which he presented his theory of knowledge *The Philosophy of Rhetoric* (1936).

That book is best remembered for its discussion of metaphor, which

provided a focus for the study of "how words work," Richards' insou-
ciant definition of the new rhetoric for which he was calling. He also
developed there what he called his "contextual theorem of meaning."
The concentration on language did not mean that he forsook an interest in mind considered as "the nervous system," but there is a difference: It is now a matter of the *philosophy* of perception. Richards knew
better than any critic of his time that "the prime agent of all human
perception" provides the forms by whose means we perceive and that
perception is a model of all form-creating and form-finding, of both
composing and recomposing, as he called reading. "The initial terms,"
he wrote of his theorem, "are not impressions; they are sortings, recognitions, laws of response, recurrences of life behaviors" (p. 30). His
discussion of primordial abstractness and generality is a careful audit of
what it means to speak of imagination as "the shaping spirit." The
study of perception is important to the philosophy of rhetoric because
it provides the motive for a theory of comprehending: When we read,
our conclusions about what the author meant "rest upon analogies—
certain very broad similarities in structure between minds." The fundamental law of mind—the all-in-each of every mind—seems to be that
"the simplest scrap or pulse of learning and the grandest flight of
speculation share a common pattern."[15]

Thus, appetencies and impulses give way to mind and imagination,
which, are, of course, philosophical concepts. Richards explicitly recognizes the inadequacies of psychology by declaring in *The Philosophy of
Rhetoric* that rhetoric, which is an inquiry into "how words work," must
be philosophic:

> It must take charge of the criticism of its own assumptions and not accept
> them, more than it can help, ready-made from other studies. How words
> mean is not a question to which we can safely accept an answer either as an
> inheritance from common sense, that curious growth, or as something
> vouched for by another science, by psychology, say—since other sciences
> use words themselves and not least delusively when they address them-
> selves to these questions.[16]

The farther he ventured toward a philosophy of mind, the greater the
distance he put between himself and that earlier "science of symbolism"
first explored in *The Meaning of Meaning*. It was about this time that
Ogden and Richards began to misunderstand one another.

Richards had gone to China in order to establish Basic English as a
medium of cultural exchange. He held the view that China, soon to
emerge from her ancient isolation, would need to have means of access
to the ideas of the West just as we would need a means of understanding Chinese life and history. The world community needed not a world

language (which is the way Ogden thought of Basic English) but a channel for transmitting ideas and values from one language to another. The Rockefeller Foundation, which had underwritten the China years, now urged Richards to go to Harvard, where he would be able to develop primers in Basic English, using pictorial representation of what was being said. Thus, Richards came to Harvard in 1939 as lecturer in education and director of the Commission on English Language Studies. He was based at the Harvard Graduate School of Education, but he was appointed university professor in 1944, and, until his retirement in 1963, he lectured and offered courses universitywide.

As I have been claiming all along, the practical was for Richards never separable from the theoretical: In moving from one Cambridge to the other, his further aim was to make room for mind and imagination at the center of educational design. His ambition, when he first arrived at Harvard, was seemingly boundless. The discoveries set forth in *Practical Criticism* had convinced him that what was needed was a method of teaching reading, but throughout the 1930s the conviction had grown that that method had to be grounded in *elementary* education. Richards, of course, took the ambiguity to heart, turning his attention to ways of changing how children were taught to read. He also concerned himself with the way teachers were taught to teach and with the reform of the curriculum of higher education. He was an active member of the committee that prepared *General Education in a Free Society* (1952), the report that established "Gen Ed" at Harvard, a prototype of subsequent core curricula. Meanwhile, he was producing, with Christine Gibson, the series of primers, of which *English Through Pictures* (1952) was the model. His aim in later years was to design an escape from world catastrophe by means of global literacy in a campaign to be waged with the help of Basic English and electronics.

It was Coleridge, he said, who led him "from criticism to creation," but he soon discovered that his literary colleagues, both in the United Kingdom and in the United States, considered his new undertaking futile and demeaning. Richards put it this way to Reuben Brower:

> Do you know when I decided to back out of literature, as a subject, completely, and go into elementary education, I learnt something. I learnt where the academic railway tracks are. I was crossing the railway tracks in the most sinister fashion. I was told so again and again. . . . In a way you are betraying a cause, showing things up [Brower, pp. 29–30].

Interpretation in Teaching (1938)[17] was to have done for the reading of prose what *Practical Criticism* had done for the reading of poetry; it had no such effect. When Brower asked Richards if it had been "the grand hinge from one way to another," Richards answered, "From things which had been strangely successful to things very much otherwise."[18]

It must be said that *Interpretation in Teaching* is crotchety and ill-written, sounding more like a poorly transcribed interior monologue than a cogent presentation of either theory or practice. Nevertheless, Richards considered it one of his best books, probably because he thought he had demonstrated just how deep educational reform would have to go if *remediation* was to be successful—and he was dismayed by neither the term nor the concept. *How to Read a Page* (1942)—a much better book, in my opinion—followed shortly; it was Richards' response to Mortimer J. Adler's *How to Read a Book*. Whereas the Chicago educator offered useful advice on how to skim and summarize, how to use the end papers for your own index, Richards returned to the techniques of multiple definition and interpretive paraphrase, showing how they might be deployed in interpreting, not a Chinese sage this time, but Whitehead and Collingwood. He juxtaposed a standard translation of a passage from the *Posterior Analytics* with one in Basic English, to demonstrate the principle that ambiguities are "the hinges of all thought." He listed "a hundred great words," claiming that because they allow us to make distinctions and connections, we couldn't think without them. Throughout, his aim was "not . . . improved theory of language but improved conduct with it" (p. 47).

He wanted readers to *slow down*—Richards was the antitype of Evelyn Wood—so that they could discover how questioning is related to defining; how interpretation entails at every step an active construing of the interdependencies of meanings. There is scarcely an issue or concern of current rhetorical theory and practice that would not be illuminated by a study of this book. By being a handbook for interpretation, it provides "a course in efficient reading"—Richards was one of the inventors of the self-help book—and, because readers must continually set down responses and paraphrases and queries, the book provides as well an excellent course in writing, seen as the making of meaning. It is, furthermore, a guide to an understanding of the moral dimension of all we do with language: Richards concludes a preface to the second edition of *How to Read a Page* by declaring that the book's "ultimate theme is Purpose; its own purpose being to offer, through a clearer eye for what we do as we think, a juster position for living." This passionate, provocative, and instructive book is unfortunately the least well-organized of Richards' books, and it suffers the most from oblique animadversions and disconcerting digressions. It provides, nevertheless, a foundation for the theories of reading, of interpretation, of composition, of comprehending that Richards was to generate in subsequent books, now mostly forgotten.

His sharpened sense of the need for philosophical guidance in developing a science of symbolism came, I think, from Richards' discovering over and over the practical requirements as well as the theoretical implications of certain ideas that had been important to him from the first, chiefly Peirce's theory of the sign. The Peircean sign is triadic: It is constituted by a *representamen* (or symbol), an *object* (or *referent*, in the Ogden and Richards terminology), and an *interpretant* (or *reference*).[19] The only way to get from symbol to what is symbolized is by means of a mediating idea, which must, in turn, be interpreted. When Peirce declared that each sign requires another for its interpretation, he was showing that the interpretant of one sign becomes the representamen of the next. Richard's recognition of the importance of this interdependence of symbol and meaning (or meaning-making) is reflected in his circular formulations (*thinking* he defined as "arranging our techniques for arranging"), as well as in his emphasis on the fact that all critical inquiry into functions of, and factors in, language must be conducted by means of language. An absolutely central principle of Richards' philosophy of rhetoric is that we think not just *about* concepts but *with* them. These necessary interpretants of all symbolic relationships he called "speculative instruments."

To recognize the significance of mediation is simultaneously to understand representation and interpretation or, rather, their relationship: to understand that comprehension is a matter of judging *what is said* and *what is meant* as mutually constraining functions. Since all import is mediated, we must explore contexts and perspectives, situations and purposes: Language and thought are never to be considered separately. This essential dialectic of language and thought Richards called "the continuing audit of meaning," delighting, no doubt, in the pun. To begin, as Vygotsky has it, with "the unit of meaning"[20] generates questions of great interest to both the psychologist seeking to understand the mind's powers and to the teacher who wants to guide them. When the emphasis is on "the continuing audit of meaning," there is every chance to exercise and to learn control of the reflexive capacities of language; for Richards there never was any danger that this would become, as it has in certain circles today, a narrow, spirit-killing interest in language about language, with never a concern shown for *purposing*.

For the dialectician, beginning with meaning entails recognition of the fact that we cannot get under the net of language; the correlative is the discovery that language is not simply a medium but a means. Because we can have no direct, *im*mediate (unmediated) knowledge of the world, we cannot claim absolute truth for our statements; we must, therefore, cultivate what Peirce called "a contrite fallibilism." I do not think that Richards is ever contrite, but the idea of keeping things tentative is at the heart of his pedagogy. As nothing in the triadic relationship is stable, we must learn to take advantage of the fact. Only in the recognition of the open-endedness of the process of comprehending can the learner—the reader, the interpreter, the writer—discover the heuristic powers of language itself.

There are many barriers to that understanding of language, but the chief one Richards saw as the misuse of information (or communication) theory. The diagram used by information theorists and computer engineers represents how the "signal" is "coded" so that "information"—defined as absence of "noise" in the "channel"—can be assured a safe passage until it is "decoded." But that is not the same thing as representing how a message is conceived or understood; *signal* and *message* are continually confused by rhetoricians deploying the terms of communication theory. The Morse Code, Richards remarked, is not of a kind with a code of behavior, but that differentiation is lost on rhetoricians who depend on Roman Jakobson's model of the communication situation. Like all positivist models, it begs the question of what exactly is being represented. A model is a form by which we represent to ourselves how a process takes place, how something operates; positivist models cannot represent unquantifiable processes like the making of meaning or the composing process because their designers reject the very concepts at issue or rename them so that they appear manipulable; for example, language becomes "verbal behavior." The confusions and inadequacies of current rhetorical theory stem from the fact that positivist models are employed to explain functions that have not been conceptualized, if indeed they have been recognized. What we are most frequently offered is really an up-to-date version of the ancient view of language as the garment of thought. Richards named it the "Vulgar Packaging View" and had this to say about it:

> What's chiefly wrong with it? This. It stands squarely in the way of our practical understanding and command of language. It hides from us both how we may learn to speak (and write) better, and how we may learn to comprehend more comprehensively. Managing the variable connections between words and what they mean: what they might mean, can't mean, and should mean—*that*—not as a theoretical study only or chiefly, but as a matter of actual control—that is the technique of poetry. If anyone is led into a way of thinking—a way of proceeding, rather—as though *composing* were a sort of catching a nonverbal butterfly in a verbal butterfly net, as though comprehending were a releasing of the said butterfly from the net, then he is deprived of the very thing that could help him: exercise in comparing the various equivalencies of different words and phrases, their interdependencies, in varying situations [*So Much Nearer* p. 175].

If rhetoricians read such essays as "The Future of Poetry," from which this passage comes (in *So Much Nearer*, 1960) and "Poetry as an Instrument of Research" (in *Speculative Instruments*, 1955), they would find reasons both for resisting the seductions of psycholinguistics with its scientistic lingo and its fraudulent "models" and for returning to what they might know about the interinanimation of words from their study of literature.

Richards patiently and wittily demolished Roman Jakobson's wiring diagram, reassembling it to show how it could be made to represent context fields, for "messages are generated by context."[21] In other diagrams and models, Richards tried to show how *purposing* is at the center of all meaning-making; how sentences function in situations (*sens/sits*); how *spokens* and *meants* and *writtens* are related. None of these concerns can be represented in the most popular of all positivist diagrams, the ubiquitous "triangle of discourse," which has helped to institutionalize confusion. The logic of the three points—encoder/writer, message/text, decoder/audience—is obscure because there is no way of representing the dialectical relationships of the three points, nor is any place found for intention or meaning or interpretation. Most "innovative strategies" and new "paradigms" are created simply by finding new names for the points of this triangle, new ways of differentiating them, though with respect to *what* is not always clear. And that is the point: the "triangle of discourse" is a triangle without *triadicity*.

For Richards, pedagogy was dialectical, or it was nothing. He continually argued that as we determine *what* it is we are trying to do, we will learn *how* to do it. That does not mean that pedagogy is a matter of first deciding aims and goals and then setting the dials and letting the teaching proceed automatically, nor is it a matter of proceeding aimlessly. Deciding what we are doing is not an exercise in problem solving but in *problematizing*, to use an ugly term associated with Paulo Freire. The *how* continually provides feedback so that we reformulate the *what;* the emerging intention provides feedforward. Like Kenneth Burke, Richards knew that forming is dialectical, an act whereby questions are framed in accordance with expected answers; and he knew, of course, that because composing is forming, it, too, is dialectical.

The distinction between process and product has its uses, but it has become institutionalized in current rhetorical theory and composition pedagogy as a dichotomy, and as such it begs most of the questions we should be concerned with. Here is a description that could help us return to a dialectical conception of what we do when we write:

> Composition is the supplying at the right time and place of whatever the developing meaning then and there requires. It is the cooperation with the rest in preparing for what is to come and completing what has preceded. It is more than this, though; it is the exploration of what is to come and of how it should be prepared for, and it is the further examination of what has preceded and of how it may be amended and completed [*So Much Nearer*, pp. 119–120].

Richards' pedagogical inventions are all conceived of as ways and means of ensuring that interpretation is in the control of the meaning-maker who poses questions in order to determine what the choices are, choices that, in turn, determine the meaning. The process of interpretation is never stable or linear; it is not a matter of solving problems but of posing them in the light of tentative solutions. Richards saw all discourse as a state where interdependencies of meanings were ordered and controlled. But it is a democratic state in which that order was always subject to the voice of emergent meaning—the constituency with the power. The reason that poetry can serve us as an "instrument of research"[22] is that it offers the most fruitful opportunity to study this making of meaning.

All writing Richards considered as work in progress; the protocol, the best-known of his pedagogical inventions, he regarded as the representation of a response, a record or account of a process of arriving at saying what was meant. No one is likely to improve on the following caveat issued in *Interpretation in Teaching* on the matter of interpreting protocols:

> These scraps of scribble are no more than faint and inperfect indications—distant and distorted rumors—of the fleeting processes of interpretation we are trying to study. They are never to be read *by the letter* (another of the tired pedagogue's besetting sins); they do not tell their own story; they are mere clues for us to place and interpret in our turn. What they indicate are phrases, moments, slices or sections—abstractly registered and perpetuated for our inspection—of processes which were on their way before the pen walked on paper, processes which went on in semi-independence while it walked and afterwards. We have to remember, unless we are to forget all that we have to teach, that what the writer meant is not to be simply equated with what he wrote [p. 30].

Derived from his interpretation of protocols and from his theory of translation, the technique of multiple definition allows fresh approaches to the "problem" of authorial intention, which is tirelessly addressed in the classroom with the single, undialectical question, "What is the author trying to say?" Multiple definition proceeds by reduction and expansion, by asking such questions as, How does it *change* the meaning if we put it this way? What *difference* does it make if we reverse these lines? These assertions? If we read A as a noun, doesn't that bring B into sharper focus? If we construe X in this way, can we then hold onto the earlier reading of Y? Used as a means of reviewing and revising protocols, multiple definition is an excellent bridge from reading to writing—and back again. Multiple definition and interpretive paraphrase can help us rethink the relationship of reading and writing and to understand the implications of claiming that interpretation is central to both.

A concern for actual practice provides the context in which we can understand what seems otherwise an obsessive concern on Richards' part with Basic English. Critics have almost universally derided his continual setting forth of Basic English, which they have called "barbaric" or foolish. They have generally not considered what he was urging: Basic English is a "technique for language control," a means of reducing and expanding texts in the interest of generating multiple definitions. By experimenting with it in the classroom, rhetoricians could discover what some reading experts and ESL teachers have long known, that Basic English is a technique to assist in the making of meaning by activating choice.

If Basic English has met with scorn, Richards' "specialized quotation marks" have been received in an embarrassed silence. They made their first appearance in *How to Read a Page* in a note appended to a chapter in which Richards had carried out a critical examination of critical assertions about critical matters. Of course, he delights in this circularity and foregrounds it by little marks placed around one and another word or phrase, signaling the sense in which the word is to be taken. Those marks play an important role in the centerpiece of *Speculative Instruments*, "Toward a Theory of Comprehending," where he comments as follows:

> Once we recognize to what an extent thinking is a taking of account of the conduct of our words, the need for a notation with which to study and control their resourcefulness becomes obvious [p. 30].

If impatience can be contained, if the reader is willing to slow down the rate of reading in order to study the process of interpretation, these little shrieks and queries and labels can be as instructive as page-long commentaries. William Empson, in *The Structure of Complex Words*, warned that "if you jam the literary criticism and the linguistics together, you interfere with the normal processes of jugment." Richards apparently believed that, in teaching critical reading, concatenation is precisely what can help develop those processes. He is concerned to follow as closely as possible the dynamics of concept formation, and that requires some way of signaling shifts from the particular to the general, from the general to the particular. Thus, the chief use of the specialized quotation marks (which he later renamed "meta-semantic markers," in a parody of linguistic lingo) is to demonstrate the full significance of Peirce's differentiation of *type* and *token*, to illuminate the consequences that differentiation has in developing an understanding of "how words work."

Richards' protocols and his techniques for multiple definition, the reductions and expansions of interpretive paraphrase and "translation" into Basic English, the meta-semantic markers and the fantastic diagrams—all are ways to bring theory to the test. They are meant to help us define context; ascertain field and range of reference; establish perspectives. They alert us to that mysterious process whereby meanings are formed and transformed. All are deployed in the service of dialectic, the "continuing audit of meaning." It is, nevertheless, true that none of these "techniques for language control" has had any appreciable effect on the way anybody teaches reading and writing. They remain idiosyncratic—unadaptable, cumbersome, capable of producing effects exactly opposite of those intended. It is not the techniques but the principles they are meant to realize that are the more likely to change our practice, if we can get at them. It is my conviction that by searching the entire range of Richards' thinking about language and learning, we can find the principles we need to guide our teaching of writing just as a generation of teachers found ways of teaching close reading from studying Richards' early books. Let me suggest what these principles might be.

Richards believed, from his earliest years as teacher and theoretician, that theory and practice must be kept together. He insisted on calling himself a "linguistic engineer." Late in his life he commented:

> The Principles [of a design of instruction] have governed the choice of detail. . . . Unless the principles stay constant, fertile interplay between what is looked for (feedforward) and what actually happens (feedback) is precluded. The experimentation will not lead to the strengthening or weakening or emendation of the principles which should be its main purpose [*Complementarities*, p. 249].

Keeping the principles "constant" but subject to emendation is not the trivialized pragmatism that gives us "behavioral objectives" but a Peircean pragmaticism in which a tentative aim is set over against the emergent actual so that "design of instruction [can be] a self-governing field of study."[23]

To assure the dialectic of feedback and feedforward, Richards insisted that assignments should form a sequence, with reading and writing always together. The challenge is to arrange matters so that there are not too many problems at once; so that the learner can "see what he's doing" and can, when "the new, partially parallel task" comes along, know "what is required of him and how he can meet it." Keeping the complex simple is no paradox, if complexity is a matter of "meaning-charged language" and simplicity, the character of the presentation. Richards' plea is for "organic" sequences: "What follows depends on what has come before and in turn protects, confirms it, and illuminates it" (*Speculative Instruments*, pp. 96, 97). Once the mind of the learner is engaged, there will be no need for that "adventitious jollying up" that characterizes conventional education design; when the mind is engaged, the learner will require no further motivation; he or she will move along, drawn by "the lure of the task itself."[24]

Another fundamental principle is that from the scrutiny of the *what*, the *how* can be derived. Here is one of many formulations of this conception of method; what we need, Richards wrote in *Design for Escape* (1968), is

> sets of sequenced exercises through which . . . people could explore, *for themselves*, their own abilities and grow in capacity, practical and intelligential, as a result. In most cases, perhaps, this amounts to offering them *assisted invitations* to attempt to find out just what they are trying to do and thereby how to do it.[25]

Richards thought that programmed instruction, especially if it could be computerized, would be the best way to clear out the distracting problems, that binary oppositions could model certain of the choices necessary in construing, but that they were never to be divorced from the purpose of making evident the choices entailed in working with "meaning-charged language."

Entailed in the scrutiny of the *what* is conscious and deliberate activity: Richards was absolutely committed to the idea of the heuristic value of thinking about thinking. Making conscious what anybody does "naturally" with language is an absurd idea for those who think of thinking as a matter of association or data retrieval and of language as "verbal behavior." What Richards meant by thinking about thinking was not that the teacher should say to the student, "Now you must realize that what you're doing here is called *generalizing*." He meant, rather, that the organic sequence itself enables the learner to think about his or her thinking as a means of making meaning. In the course of his invaluable discussion of metaphor in *The Philosophy of Rhetoric*, Richards writes as follows:

> *Thought* is metaphoric, and proceeds by comparison, and the metaphors of language derive therefrom. To improve the theory of metaphor we must remember this. And the method is to take more note of the skill in thought which we already possess and are intermittently aware of already. We must translate more of our skill into discussable science. Reflect better upon what we do already so cleverly. Raise our implicit recognitions into explicit distinctions [pp. 94–95].

Those "implicit recognitions" are the work of the imagination, "the prime agent of all human perception." Richards' pedagogy is informed always with the idea that looking and listening could together serve as a model for all that we do in reflecting to some purpose. His lifelong interest in "audio-visual aids" stemmed from the conviction that they were "aids to reflection," another Coleridgean phrase. As for those "explicit distinctions," all his teaching is aimed at showing how differences make a difference. Theories are based on them; practice requires them. "To notice what varies with what" he defined as the essence of scientific method.[26] His theory of metaphor led him directly to the concept of *speculative instruments,* those ideas we think with. The wonderful phrase is both an emblem of the ancient wisdom that seeing is a form of knowing and a reminder of the fact that science, which would be impossible without instruments, is *knowledge,* in the root sense.

These principles of course design and pedagogy are certainly at odds with the conventional wisdom of the tired old rhetoric Richards sought to replace with a new or "revived" rhetoric, as he put it in 1936, but they are also fundamentally different from the central ideas of what is currently known as the New Rhetoric. (This is expectable, for Richards' rhetoric springs from a philosophy of language different in all respects from the views underlying such techniques as tagmemic heuristics and "problem solving," which are derived from a fundamentally positivist linguistics.[27]) His points of departure, his continually inventive ways of "arranging the techniques for arranging" are full of interest for any teacher willing to try to teach reading and writing as interpretation, but the perspectives Richards' work offers are not easily won. If his theory of comprehending is not brought to the test of practice, it cannot help us; furthermore, we will need more intelligent conceptions of "testing" than those generally accepted. The smug declaration that research in the teaching of English should be undertaken for its own sake or that it should be modeled on the "mission-oriented" projects of the natural and social sciences are views antithetical to Richards' view of the dialectic of theory and practice in the realm of language. The teacher undertaking to make interpretation central will necessarily be a researcher, his or her classroom necessarily becoming a "philosophic laboratory."

In some cases, it will be difficult to enjoy the advantage Richards' ideas and methods can afford because they might seem familiar or even hackneyed. Consider, for example, the idea of paraphrasing, which for Richards provided the chief means of comparing differences, the central act of mind in interpreting. Paraphrasing does not mean substituting one word for another; the point needs to be made explicitly, now that experience of translation on the part of students and teachers alike cannot be taken for granted. Nor does paraphrasing mean creating a rival poem, the reader constrained only by the requirements of her "identity-theme." Glossing and paraphrasing in the light of Richards' theory of comprehending are attempts—necessarily partial, properly tentative—to represent *what is meant* by comparing the paraphrase with *what is said.*[28] (Thus, Basic English is an analytical language, a means of discovering and identifying purposes by testing provisional representation against the text.) This method of studying "how words work" is the antithesis of contemporary "discourse analysis," which tends to set aside meaning in the interest of identifying cohesive devices. Paraphrasing in Richards' style allows readers and writers to audit the meanings they are making; it encourages us to keep things tentative, to tolerate ambiguities so that they can serve as "the hinges of thought"; it helps us to learn the uses of provisional statement and thereby practice the skill of inference.

To take advantage of what Richards has to offer requires, then, a willingness to do what he says, to practice what he preaches, interpreting what happens in the light of new pedagogical purposes. We will need to be alert to rather special uses of familiar terms and, it must be said, we will need to learn tolerance of an eccentric, highly personal style if we are to gain access to the theory in the first place. Reading Richards can, indeed, provide excellent practice in seeing "how words work," but only if we are willing to be patient. He is an incomparably interesting guide in the search for speculative instruments, our means of making meaning in all investigations, philosophical and otherwise, right up to "the endless arch-inquiry: What are we and what are we trying to become?" (*Speculative Instruments,* p. 152). If he is difficult to read, it is because he seldom forgets the arch-inquiry, even when engaged in the most delicate interpretive operations. Richards is, in my opinion, not a great essayist. His eyes are on too many issues; digression is the most typical feature of his style; proliferating ambiguities are analyzed endlessly. Instead of preparing errata slips, he would write another book or, at least, supply a second edition with notes in which misstatements and misconceptions were diagnosed and amended. In this way, he was true to his own deepest conviction that what is needed is not disputation but dialectic, the continuing audit of meaning. His refusal to sneak under the net with absolute assertions or mystical pronouncements, his energetic enjoyment of the circularity of all knowledge exhaust us. But Richards is a master of the paragraph and of the paragraph sequence, the episode. He is a superb aphorist, and who can rival the brilliance of his quotations? Richards won't let us off the hook: He never tires of asking *what* in order to explore *how.* And he never is uninterested in *why* and *how come* and *wherefore.* His method insistently keeps the processes of articulation in the foreground because "clear consciousness of what we are doing is our best means of control" (*Mencius,* pp. 128–129). He recognized the hazards, of course, and illustrated what can happen with "too sudden an extension of consciousness" by quoting the rhyme about the centipede:

> The centipede was happy—quite!
> Until the toad in fun
> Said, 'Pray, which leg moves after which?'
> This raised her doubts to such a pitch
> She fell exhausted in the ditch
> Not knowing how to run!

Characteristically, he then observes: "But there could hardly be any advantages if there were no accompanying dangers."

Being an Evans, I am always delighted to read characterizations by his old friends that stress a Welsh quality in Ivor Richards, but I don't myself find in him the stereotypical Welsh attributes. He was not disputatious; he disguised his evangelical message with argument; his casuistries were never sobersided; in his lectures, he did not indulge in the *hwyl*—though he could leap on the tabletop to illustrate a point about balance. I don't know if he could sing, but he did read poetry more beautifully than anyone on either side of the Atlantic. In any case, Richards' teaching style went out of fashion long before he retired. In the 1940s, when the "Gen Ed" courses were new, his spellbinding became legendary, as it had twenty years before at Cambridge. The interest among upperclassmen and graduate students—about fifty in a class—was intense: they had the sense that Richards was handing down Western civilization to them, in the very act of reading. Eccentric, gnomic utterance was found not only tolerable but delightful, especially because it alternated with astonishing demonstrations. "We must look at this carefully; we must look all *around* it," he would declare, lifting his chair while still seated and walking in a tight circle, coming to rest in front of his open book. With 500 people in the required humanities course, this manner was not appropriate, nor was his assumption that he could count on a certain amount of "background" any longer a sound one.

The observation has been made that Richards was not very good at "getting down to his reader's level or in judging by just what amount he should be above them."[29] The same could be said for his pedagogy, but it was never part of his purpose to meet students on their own ground. In a lecture to graduate students I heard in the late forties, Richards observed that if people understood completely what one was talking about, that would be evidence that one had taught them nothing. His metaphor for the teacher's obligation was drawn from mountaineering: We should teach to the top of the class because otherwise we would surely lose them and probably the middle third, too—and we might not reach the bottom third in any case. The point was that, in the ascent toward learning, everybody needed to feel the tension of the rope. It is a pedagogy that can be very dangerous if the leader isn't expert at holding the rope or if for any reason those who follow lose their trust. Nevertheless, for the spellbound, that idiosyncratic style and the hazardous pedagogy were powerful influences and, together with the principles of practical criticism developed in the early books, they are still felt, especially by those who worked with Richards at Harvard in the 1940s and early 1950s and, of course, by those whom they have taught to teach.

The chief instance of Richards' influence on the teaching of writing is the famous course at Amherst.[30] It grew out of reading and writing courses designed during World War II by Theodore Baird, Reuben Brower, and G. Armor Craig. The influence of Richards (he had been Brower's tutor at Cambridge) is discernible in several features of this course, namely, the emphasis on observation and careful description as a point of departure; the sequencing of assignments as a series of "partially parallel tasks"; the gamelike character of the assignments, which helped students develop the habits of questioning assumptions and premises; the use of diagrams, drawings, and other schematic representations; and, most important, a dependence, revealed as the sequence of assignments unfolded, on the concept of analogy. (The Amherst Mafia, teachers and students now dispersed all over the country, can be identified by their use of the term *metaphor* to cover all representations of what Richards called *sames* and *differents*.) There was a persistent—indeed, maddening—emphasis on looking at the *what* of the various activities: *What did you do when you looked at the table? What did you do when you measured the table top?*[31] But in my view, such exercises were not designed to—and did not, in fact—encourage thinking about thinking. They stemmed from conceptions of language fundamentally different from those Richards held: the Amherst course was underwritten, not by a Peircean semiotics, but by a positivist operationalism. The guiding conception of language as a map for a territory of reality came not from *The Meaning of Meaning* but from General Semantics: it was Korzybski's vulgarization of Wittgenstein's early picture theory.[32]

Perhaps the most important aspect of the course was the attitude toward the papers. The emphasis was entirely on *what* had been said, with the instructor trying to elicit by marginal queries and final questions further statements of what the student meant. The *how*—grammar, organization, or whatever constitutes 90 percent of composition courses, now as then—was ignored. The paper was treated as a protocol to be revised by means of interpretive paraphrase and multiple definition. The assumption was that the *how* would come right as the *what* was carefully articulated.

Contemporary interest in the reader's response to "the text" and in writing as a process of making meaning holds pedagogical promise if it leads to a theory of knowledge, a philosophy of mind. Composition pedagogy needs far stronger support than it has had from such trivial notions as "cognitive style" and "thinking skills," contributed by modern psychology. The influence of Richards will, I think, begin to be felt anew when lively young teachers, as well as their older colleagues who have quelled cynicism, come to see that in teaching reading and writing we must be guided by a theory of comprehending. There could be no better guide to an understanding of what it means to call writing a "mode of learning" than I. A. Richards, and, as for "writing across the curriculum," surely no recent writer can offer readier means of understanding that "all studies are language studies, concerned with the speculative instruments they employ" (*Speculative Instruments*, pp. 115–116).

The attitude that assuages the occasional despair we may feel as we confront the higher illiteracy may simply be that teaching is still possible. Richards himself was never content with such a minimalist position. William Empson wrote of him as he neared eighty: "In his moderate rueful way, [he] has gone on being fertile in proposing steps forward; but he has never lost the feeling that they are minor ones, because an immense opportunity lies unrecognized just beyond our grasp. This indeed is what is so plainly lacking from our present leaders in linguistics. . . . Absence of vision is not inherently scientific; and in Richards the vision itself is the spell of the he casts" (Brower, p. 74). It is vision we need, surely, if we are to reclaim the principles of practical criticism and of a philosophy of rhetoric, in order to bring them to bear on the teaching of writing.

Notes

1. *Philosophy of Rhetoric*, p. 86.
2. This formulation I owe to James Slevin.
3. *I. A. Richards: Essays in His Honor*, ed. Reuben Brower, Helen Vendler, John Hollander (New York: Oxford University Press, 1973), p. 72. Hereafter cited as "Brower."
4. See Raymond Williams, "Cambridge English and Beyond," *London Review of Books*, 7–20 July 1983.
5. Brower, pp. 48–49. The recollection is Joan Bennett's. Richards remarks in an interview with John Paul Russo that he copied G. E. Moore's manner in the lecture hall and that he later dropped these habits (*Complementarities*, ed. John Paul Russo [Cambridge: Harvard University Press, 1967], p. 258). Those who heard him lecture in later years might not agree on this score. See Helen Vendler's affectionate account in *Masters: Portraits of Great Teachers*, ed. Joseph Epstein (New York: Basic Books, 1981).
6. Brower, p. 23.
7. "All living use of language . . . depends upon the user's discernment of how what is being said differs significantly from other things that might be said instead" (Richards' introduction to the reprint of Ogden's 1932 book, *Opposition* [Bloomington: Indiana University Press, 1967]). Richards saw opposition as "chief among the essential principles by which language works." It is as well the chief principle by which perception works—which is why opposition engages the mind of the learner.
8. Basic English consists of 850 words in five categories: words for *operations; things*, general and picturable; *qualities* and *opposites*. It was a by-product of the work Ogden and Richards did on definition in the course of writing *The Meaning of Meaning*. The questions they posed were these: What are

the words that are most useful? How far are certain words in English able to take over the work of others? The discussion I find the most interesting is in "Towards World English," the concluding chapter of *So Much Nearer*, but Richards wrote continually about Basic English. See also "The Hammer's Ring," William Empson's contribution to the Brower Festschrift.

Richards never gave up his attempt to persuade people anywhere and everywhere that Basic English provided some practical solutions to the problems of illiteracy. My friend Vida Markovic recounts a story of how Richards was rather diffident in an interview she was conducting until the subject of Basic English was raised. When he discovered that Professor Markovic had some knowledge of the system and interest in its promise, Richards tried to persuade her to join him that very afternoon on a trip to Africa for some sort of Basic English project. In 1979 he traveled to Peking as an honored guest, helping to set up a pilot study for teacher training. He fell ill and, after being ministered to for some weeks by the Chinese, was flown in a special plane to Britain, where he died in Cambridge on 7 September 1979. William Empson expressed in a memorial note the hope that the Chinese had indeed been convinced about the possibilities of Basic English "so that the heroic death of Richards in their service will merely help to make welcome the agreed procedure. They could have given him no death that would have made him happier" (*London Review of Books*, 5–8 June 1980).

9. What Richards called the "symbolic" function was the referential or communicative function of language. He thought he could best justify poetry by claiming that it did indeed make statements, but of a fundamentally different sort, the poet's purpose being not to communicate facts and opinions but to express and arouse feelings. He called them "pseudo-statements."

10. Wittgenstein, who was by this time well established in Cambridge, followed the practice of the Vienna positivists, who called their "unit propositions" *Protokollsätze*, each corresponding to a single sensation or fact. (For an excellent account of the Vienna Circle and its relationship to Wittgenstein, see Allan Janik and Stephen Toulmin, *Wittgenstein's Vienna* [New York: Simon & Schuster, 1973].) This epistemology was consonant with Bertrand Russell's logical atomism, which influenced Ogden's and Richards' conception of context.

11. Richards' experiment has often been misinterpreted, but probably never so willfully as by Terry Eagleton in a book curiously entitled *Literary Theory* (Minneapolis: University of Minnesota Press, 1983). This mischievous account discredits Richards' motive while misrepresenting his conclusions. If *Practical Criticism* is read only in the light of *Principles of Literary Criticism*, its limitations are certainly highlighted, but Eagleton did not apparently trouble to read either the introduction or the analysis that follows 150 pages of documentation. He therefore missed such statements as this: "When nature and tradition, or rather our contemporary social and economic conditions betray us, it is reasonable to reflect whether we cannot deliberately contrive artifical means of correction" (p. 301). Eagleton suf-

fers from a disease endemic among socialists: He associates careful, successful, imaginative use of language with a training that is identified with
and held to be inseparable from a detestable class structure. Because Richards labored all his life to find a philosophically sound and politically
decent foundation for education, the kind of misrepresentation in Eagleton's account is especially offensive.

12. Richards had wanted to find the Chinese Plato, to make accessible the chief
ideas of a great thinker of the East. He soon concluded that Mencius'
thought was accessible only through the way he used the Chinese language. *Mencius on the Mind* is less about the wisdom of the Chinese sage
than it is about translation and its relation to a theory of knowledge.

13. Brower, p. 33. There is an oblique allusion here to the fact that Owen
Barfield had written a study called *What Coleridge Thought*, in which he
scolds Richards for not having noted a certain argument. Barfield and
Richards managed to annoy one another, whatever the topic; neither could
resist taunting or chiding the other.

14. *Practical Criticism,* p. 208. This is the informing principle of Austin's "ordinary language" philosophy. Richards differentiated many orders of ordinary language, using Basic English as a translation medium and showing
how the search for contexts, placing sentences in situations, is the proper
method for development of a philosophy of rhetoric. I have discussed the
role of the idea of context in Richards' thinking in "I. A. Richards and the
Audit of Meaning," *New Literary History,* 14 (Autumn 1982), pp. 63–79.

15. *How to Read a Page,* p. 106.

16. *Philosophy of Rhetoric,* p. 23. One reason that the book with the curious title
The Philosophy of Composition was superannuated before it was published is
that its author, E. D. Hirsch, Jr., reversed this admonition, declaring that
rhetoric must seek guidance from psychology, that researchers in composition pedagogy must model their projects after those "mission-oriented"
ones designed by the National Science Foundation, that only psycholin-
guists can help us teach for readability, and so on. Professor Hirsch has
recently admitted that he was wrong on this score, but his new tack—he
wants the schools to instill "cultural literacy"—is no more promising. Like
most other researchers in the teaching of English, Hirsch could profitably
study the principles set forth in *Mencius* concerning the need for the practical criticism of our hypotheses.

17. Although *Interpretation in Teaching* was published two years after *The Philosophy of Rhetoric,* it was, in fact, written before. When Richards was invited
to give the Flexner Lectures at Bryn Mawr College, he realized that the
work he was finishing was inappropriate. He wrote *The Philosophy of Rhetoric* as a theoretical gloss on the discussions of the practical problems dealt
with in *Interpretation of Teaching,* which he thought of as the more significant book.

18. Brower, p. 30. It is notable that the Festschrift Brower edited with John
Hollander and Helen Vendler includes only a single contribution, from an
English schoolmaster, on Richards' educational theory and practice.

19. As a "sop to Cerberus," Peirce often identified the *interpretant* with the

interpreter, but it is a *logical*, not a *psychological* concept: there is no "empty sign" in Peircean semiotics.

20. *Thought and Language* (Cambridge: MIT Press, 1962), p. 80. It would be instructive for teachers of composition to compare the account Vygotsky gives of concept formation as a "movement of thought . . . constantly alternating between the two directions, from the particular to the general, and from the general to the particular" with Richards' conception of "the continuing audit of meaning."

21. "Functions of and Factors in Language," in *Poetries: Their Media and Ends,* ed. Trevor Eaton (The Hague: Mouton, 1974). See also "Towards a Theory of Comprehending," in *Speculative Instruments (SI)* and the essays collected in *Design for Escape.*

It is sometimes said that Richards came under the "influence" of information theory. I have commented (in "I. A. Richards and the Audit of Meaning") on his frequent use of the terminology of audiovisual technology, cybernetics, computer programming, and so on. Another explanation is offered by Kathleen Raine, who writes in her autobiography as follows:

> Ivor is a critic who, falling in love with the texts he studied, took to poetry; a splendid example to set against those poets who, led astray by the magpie criticism, became critics. I remember a lecture he gave . . . on the "Ode to the West Wind," ingeniously illustrated . . . with little drawings on the blackboard of electric wires and switches and boxes, meant to represent "communication" from, as he said, an unknown source, to an unknown recipient: a process beginning and ending in mystery. Shakespeare would have done it with airy sprites, Blake with angels. The little diagrams were the vestiges of a style by whose disguise, in the twenties, it was necessary at least to appear to be "scientific." But the thought was metaphysical and Platonic [*The Land Unknown* (New York: George Braziller, 1975), p. 37].

22. *SI*, pp. 144–152. Richards used the same title for a different essay collected in *Poetries: Their Media and Ends.*

23. "Structure and Communication," in *Structure and Art in Science,* ed. G. Kepes (New York: George Braziller, 1965), p. 128.

24. "Structure and Communication," p. 135.

25. *Design for Escape,* p. 97. In my textbook *Forming/Thinking/Writing* I have called the exercises "assisted invitations."

26. "Structure and Communication," p. 134.

27. In "Forming Concepts and Conceptualizing Form" and elsewhere in *The Making of Meaning* (Upper Montclair, N.J.: Boynton/Cook, 1981), I have discussed the positivist epistemology of psycholinguistics and the so-called New Rhetoric.

28. A glance at the study questions of any rhetoric-reader will reveal why this skill is moribund: What students are generally offered is a list of leading questions, requiring merely that the question be rephrased as a statement; or they are asked questions that are comfortable versions of fill-in-the-blank quizzes.

29. W. H. N. Hotopf, *Language, Thought, and Comprehension* (Bloomington: Indiana University Press, 1965), p. 114n.

30. For an excellent critique of the philosophical assumptions underlying the Amherst program, see James H. Broderick, "A Study of the Freshman Composition Course at Amherst: Action, Order, and Language," *Harvard Educational Review*, 28 (Winter 1958), pp. 44–57. In several recent conversations, Professor Broderick has added greatly to my understanding of Richards' theory and practice. Broderick studied with him and served as a teaching assistant in the Humanities course at Harvard.

    Another influence is by way of what is sometimes called, in a pleonastic phrase, "epistemic rhetoric," a rhetoric that focuses on the relationship of language and knowledge. My own work has been so classified. I have found Richards' idea of dialectic as "the continuing audit of meaning" invaluable in all my teaching. His understanding of the "primordial abstraction" of perception is, I have found, entirely consonant with the psychology that underlies Cassirer's philosophy of symbolic form. The conception of "speculative instruments" has itself served as a speculative instrument for me in reading Peirce, in seeing what it means to say that mediation is necessary to all acts of mind, to all functions of language. In *The Making of Meaning*, I have explored the pedagogical implications of these ideas, and in *Forming/Thinking/Writing* I have tried to put them into practice. *Reclaiming the Imagination* offers perspectives for teachers who want to see the classroom as a "philosophic laboratory."

31. An amusing account of a young instructor coming to grips with these assignments can be found in Alison Lurie's early novel *Love and Friendship*.

32. "General Semantics" was a movement aimed at propagating and institutionalizing the views of Alfred Korzybski, a Polish nobleman who, like Ogden and Richards, believed in the importance of a "science" of symbolism. Unlike them, he had no understanding of literary meaning and no interest in pedagogy, which is not to say that some distinguished teachers have not derived certain useful approaches from Korzybski, James Moffett and Theodore Baird among them. The General Semanticists, chiefly through S. I. Hayakawa and Stuart Chase, popularized the "Ladder of Abstraction," by which they represented their view of abstraction as the opposite of reality, of "what is happening." They institutionalized the use of quotation marks—on the page and in the air, as gesture—to signal the fact that the words "apple pie" can not be eaten. It may be that Richards developed his specialized quotation marks—*How to Read a Page* appeared when General Semantics was in its heyday—as a way of suggesting that matters of intention and reference are not as simple as Korzybski thought.

    Railing continually against "two-valued Aristotelian" logic, the General Semanticists nevertheless dichotomized at every opportunity. (They completely misunderstood the concept of a "universe of discourse.") They preached that the basic trouble (everywhere) was that people confused language and reality, and they blamed everything from war to tooth decay on that confusion, which is, of course, fundamentally different from the confusion caused by failing to differentiate *what is said* from *what is meant*. Stuart Chase tried to argue that there was no such thing as "fascism"—only fascists. Barrows Dunham, in a well-known essay called "That Problems

Are Merely Verbal," and Max Black, in *Language and Philosophy*, showed how illogically Korzybski had proceeded. The fastest demolition job was reportedly performed by Willard Van Ormond Quine on a paper napkin in the course of lunch at Eliot House. Susanne K. Langer's disquisition on Stuart Chase's cat (in *Philosophy in a New Key*) is still an amusing and useful critique of the confusions of General Semantics in the field now called "semiotics."

Selected Publications of I. A. Richards

The following books, listed in order of publication, are discussed in the preceding essay. Many of them went into a second or third edition, and most were reprinted in paperback, often by other publishers. Those still in print are indicated by an asterisk (*). Bibliographical data for articles and books briefly mentioned can be found in the notes.

The Meaning of Meaning (London: Kegan Paul, 1923).
Principles of Literary Criticism (London: Kegan Paul, 1924).
**Practical Criticism* (London: Kegan Paul, 1929).
Mencius on the Mind: Experiments in Multiple Definition (London: Kegan Paul, 1932).
Coleridge on Imagination (London: Kegan Paul, 1934).
**The Philosophy of Rhetoric* (New York: Oxford University Press, 1936).
**Interpretation in Teaching* (New York: Harcourt, Brace, 1938).
How to Read a Page: A Course in Effective Reading, with an Introduction to a Hundred Great Words (New York: W. W. Norton, 1942).
English Through Pictures (New York: Pocket Books, 1952).
Speculative Instruments (Chicago: University of Chicago Press, 1955).
So Much Nearer (New York: Harcourt, Brace and World, 1968).
Design for Escape: World Education Through Modern Media (New York: Harcourt, Brace and World, 1968).
Techniques of Language Control (Rowley, Mass.: Newbury House, 1974).

A fully annotated bibliography by John Paul Russo is included in *I. A. Richards: Essays in His Honor*, ed. Reuben Brower, John Hollander, and Helen Vendler. The interview Brower conducted provides a lively account of Richards' long and complex career; the one Richards gave Russo (in *Complementarities*) includes some interesting comments on pedagogy. *The Rhetoric Society Quarterly* (Fall 1980) includes essays on Richards by T. Y. Booth, Thomas Derrick, and Janet Kotler. Francis M. Sibley has written instructively on "speculative instruments" in "How to Read I. A. Richards," *The American Scholar*, 42 (1973). Jerome P. Schiller includes an excellent bibliography of critical assessments of Richards' literary theory and criticism in his *I. A. Richards' Theory of Literature* (New Haven: Yale University Press, 1969).

Richards' work through 1942 is the subject of a book-length study, *Language, Thought, and Comprehension*, by W. H. N. Hotopf (Bloomington: Indiana University Press, 1965). Neither Richards nor any commentator I have read, other than Schiller, mentions this book, probably because it is a book about books about books about books. It is hard to read and use because it is written with a very English disregard for the need to clarify references (grammatically and substantially) as one moves from spoken to written discourse. Hotopf's aim is to reveal "the anarchic consequences of too great emphasis upon self-expression in defiance, rather than through transcendence, of convention" and to "show that, by letting his theory influence his practice, Richards revealed, both in an excessive failure to communicate in his later books and in certain confusions of his thinking, the dangers of his course." Hotopf concerns himself with analyzing the misreadings Richards has suffered at the hands of John Crowe Ransom, Max Black, and William Empson, among others; but his own critique, sober and fair-minded as it is, falls short because he is looking for consistency and system in a highly unsystematic thinker and because he has no understanding of "triadicity" (Peirce gets only a footnote in this long study) nor of mediation. Richards' insistence on the centrality of purpose and the necessary role of meanings in the making of further meanings illustrates for Hotopf "the Chinese box effect in Richards' thinking." I doubt that Hotopf would find in the nine books Richards wrote after the ones considered in this "case study" any reason to modify his charges.

Introducing Richards in later years as a visiting lecturer, a professor suggested that "one direct avenue to a truly liberal education would be to read Dr. Richards' books straight through, from the very first to the most recent." "No! No! No!" cried Richards. "You should *begin* with the most recent and then, if you wish, go back to the earlier books." For any prolific writer, the book just finished will be of greatest interest, but his or her readers cannot come to the current work with useful expectations to guide them unless they have, indeed, some knowledge of the forerunners. My own opinion is that Richards' later books should be read in conjunction with the earlier studies; in that way, we can benefit from "the continuing audit of meaning," the continual reassessment of principles of language and learning.

Richards told John Paul Russo that "Towards a Theory of Comprehending" and "Emotive Meaning Again" in *Speculative Instruments* ("what I fancy is my most intelligent book"), together with "Meanings Anew" in *So Much Nearer*, "say better what we tried to say earlier"—in *The Meaning of Meaning*—and that in reading them "you will have what I still want to propose." Writing teachers should certainly add "Some Glances at Current Linguistics" (*So Much Nearer*), which does for Chomsky what Chomsky did for B. F. Skinner. Then if the reader goes to *Mencius* and *The Philosophy of Rhetoric*, reading them dialectically with *Design for Escape* and *Techniques for Language Control* or with miscellaneous pieces in *Complementarities*, there is a good chance that he or she will discover some servicable speculative instruments.

In any case, it's important, I think, to read Richards in new and different contexts—not with Allen Tate and Monroe Beardsley but with Paulo Freire and Maria Montessori, with Jane Addams and William James—with other great teachers and philosophers of education. (See Part III of *The Making of Meaning* where I have tried to provide such a context.) Read contentiously, with an eye to system-building or proof-garnering, Richards is not very useful to us; read with hope by those who are willing to consider the classroom a "philosophic laboratory," Richards reveals that even his contradictions are heuristic. Richards' own musing suggestion is our best guide: "Read it as though it made sense and perhaps it will."

STERLING ANDRUS LEONARD

by John C. Brereton

Sterling Andrus Leonard was born in 1888 in National City, California, a suburb of San Diego. His father was a dentist; his mother, an English teacher who inspired him with a love of language and literature.[1] After his father's death he attended Simpson College in Indianola, Iowa, and transferred to the University of Michigan, from which he received a B.A. in 1908 and an M.A. in 1909. He then held a series of short appointments—at Michigan, at the Milwaukee State Normal School, at the Danzig Gymnasium in Germany in 1911–12 as an exchange teacher, at the University of Wisconsin, and at Horace Mann School in New York City. In 1920 he became assistant professor of English at the University of Wisconsin, where he remained until his death.

Leonard's contemporary reputation and influence came from publications in two separate areas, each of which marked a distinct phase of his career: from 1915 to 1926 he published largely on pedagogical issues; from 1927 until his death he was immersed in scholarly research on the history and current state of English usage. Early in his career he earned a reputation for his innovative approach to composition pedagogy, in which he advocated an almost total reliance on group work, a concern for the writing process, a strikingly modern view of audience analysis, and a strong advocacy of more rigorous composition research. His books in this field include *English Composition as a Social Problem* (1917), *Essential Principles of Teaching Reading and Literature* (1922), and a number of texts and anthologies. Today, however, these early works are largely unknown and unavailable; what present renown Leonard enjoys comes from the work he produced after 1927, when he began his substantial body of research on the issue of "correct" English. His dissertation, published as *The Doctrine of Correctness in English Usage* (1929), demonstrates how modern notions of usage evolved during the late eighteenth century; it remains the standard book on the subject. And his most influential book, *Current English Usage* (1932), is a path-breaking empirical study of opinion about English punctuation, word usage, and grammar.

Throughout his life Leonard was a busy, productive scholar, publishing seven books and over thirty articles. His professional rise was rapid; he became associate professor in 1925 before he had earned his doctorate. He was an active member of the National Council of Teachers of English (NCTE), serving on important committees; in 1926 he was president. In 1927 he took his doctorate at Columbia, working with George Philip Krapp, the historical linguist whose special study was English usage. Leonard's dissertation was published immediately, and he embarked on an ambitious study of modern usage. But Leonard was not to live to see his second book through the press. By 1931 he had completed all the preliminary studies and was at work on what would be *Current English Usage;* he looked forward to a year's leave and the visit to Madison of I. A. Richards, with whom he had long corresponded.[2] When Richards arrived, Leonard took him boating on Lake Mendota, on the Wisconsin campus. Their boat overturned, and, within sight of the university boathouse, Leonard drowned, and Richards barely survived. Leonard was forty-three.[3]

Testimony about Leonard as a person indicates that he was a memorable character. His Wisconsin colleague William Ellery Leonard described him as "ruddy and round-faced with expression playing back and forth between quizzical and grave . . . impetuous in gait, gesture, and speech." He characterized Leonard's intellectual stance as a "sane, balanced, alert, and genial radicalism" (WEL, p. 179). Another colleague, Robert Pooley, a fellow language researcher at Wisconsin and NCTE president in 1952, described him as "impulsive and dramatic." Pooley said Leonard's teaching

> was characterized by freshness of viewpoint, by quickness of mind, and by an infectious enthusiasm. The latter drew to him many disciples, who were both challenged and exasperated by the brilliance of his ideas. Leonard advanced by flashes of insight rather than by logical steps; his students found him at times a meteoric shower rather than a lodestar [Pooley, p. 1].

Such an assessment makes one suspect that Leonard was not a person for day-to-day constancy, for dogged efforts, or for seeing a tedious project through to its conclusion. And it is true that there is little of the tedious in Leonard's approach to teaching or language studies. He does not suffer fools gladly; his impatience with traditional approaches shows through often, as does his exasperation with what he calls "Old Purist

Junk" in contemporary textbooks. Yet his work on eighteenth century theories of grammar is remarkably exact and thorough. And *Current English Usage,* though completed by others after Leonard's death, bears every sign of careful preparation and thoughtful planning. Indeed, Leonard's writings show him to be a man given to thinking rigorously in categories, with a passion for classifications and a respect—perhaps exaggerated—for scientific precision.

Among composition teachers and researchers, Leonard is almost unique; he combines radical curricular innovations with solid research on language change. Most choose one field or the other, but Leonard achieved fame in both. And although Leonard's efforts were directed more toward schools than toward colleges—most of his teaching was either in high schools or in teacher training rather than in freshman composition—his legacy provides much of interest to present-day writing teachers and researchers, no matter what level they work on.

Leonard's first book, *English Composition as a Social Problem* (1917), is a thoroughgoing Deweyan approach to composition, something Dewey himself never provided.[4] The Deweyan influence is not surprising; Leonard wrote his book while teaching English at the Horace Mann School, the laboratory school of Teachers College, Columbia University, where Dewey's influence reigned supreme. Leonard's preface has a grateful acknowledgment of that influence, particularly for Dewey's notions of the socialization or group process in the teaching of writing. *English Composition as a Social Problem* is a remarkable book, stunningly modern in the way it anticipates so much of the best contemporary approach to writing. It resembles nothing so much as James Moffett's contemporary classic, *Teaching the Universe of Discourse* (1968).

Leonard devotes the bulk of the book to four issues: how students' desires to express themselves can serve to motivate them to compose; how the class can function as a group; how instructors can use group processes to facilitate learning of organization; and how to foster learning of minimum essentials of grammar, punctuation, spelling, and manuscript form.

In the writing course Leonard describes, both the composing process and the ultimate product depend on the students' motivation to communicate with other class members. He explicitly distinguishes his own approach from the artificial motives lying behind composition in the overwhelming majority of classes, where projects and topics, no matter how cleverly devised, serve mainly as pretences for teachers to check up on grammar and spelling. Leonard regards a genuine motive as the foundation for all successful teaching and distinguishes quite different motives behind oral and written composition. For oral composition Leonard lists the following:

1. The Entertainer Motive (in which students tell stories).
2. The Teacher Motive (in which a speaker instructs others or explains something he or she knows about).
3. The "Community-Worker" Motive (in which students report on investigative projects that look at social issues or needs).

Other motives, what he calls "The Presentation of Opinions" and large-scale "Projects that reach beyond the class group" (i.e., have an audience different from the group actually in the classroom), Leonard would save for specialized or advanced classes.[5] Correspondingly, the motives that lead specifically to written composition are the "Preservation of specially good work" and "Publication—reaching a wider audience." Thus, the motives for oral presentation come from the students themselves and from the material they work on; the motives for writing reside in the natural desires to preserve good work or to attract others' attention.[6]

In demonstrating how the class may function as a group, Leonard supplies practical ideas that seem commonplace today but were still revolutionary in 1917: movable chairs, discussions with students in a circle, stress on spontaneous talking before writing, emphasis on sensory detail before drawing generalizations, and group proofreading. Leonard encourages the instructor to take a backseat, to allow the group to make suggestions, and to rely on "the cooperative interest and communal pride of the class" (p. 58) for checking up and improving work.

Leonard advocates response to writing assignments by the whole class, especially because major issues like tone, inclusion of appropriate material, and interest level may best be determined by a real audience. And for individual problems, he refuses to become a typical English teacher. He would "end the regime of wholesale red ink" (p. 61) by holding conferences and by requiring students to hunt out their own errors. His own students received on their papers a number indicating the total of their errors or an asterisk next to mistakes. The bulk of the comments dealt with the thinking and the expression. (Leonard also borrows an idea from Barrett Wendell, a theme-card listing each students' strengths and weaknesses on all papers, a kind of cumulative list of students' weekly progress [p. 61].)[7]

Leonard's chapter on developing and organizing ideas completely ignores the traditional grouping according to modes of discourse: narration, description, comparison, and persuasion. Instead, he suggests a planning process, which he calls "prevision," in which students limit and organize subject matter while working together in class groups. He supplies three schemes for help in planning the most effective organization: (1) a strand of narrative, "what happened next"; (2) informal blocking into paragraphs by "throwing larger wholes into a few convenient groups"; and (3) a more highly focused organization based not on a thesis sentence but on "a grouping about an interpretive sentence which forms the writer's conclusion" (pp. 68–86). In the planning and organizing processes, Leonard insists on starting from realized experience, from details, and from close observation, and then on testing those against the class needs. He is against rigid notions of organization, particularly outlines and syllogistic logic. Similarly, his opposition to traditional groupings of ideas takes the form of a remarkably fresh insight into organization:

> Attempting to present materials under the headings 'Introduction,' 'Body,' 'Conclusion'—an apparent outgrowth of Aristotle's unassailable but unfortunately applied remark that a story must have a beginning, a middle, and an end—is, of course, no true *grouping* of the ideas, and should, I maintain, never be accepted as a substitute for one; it is the 'Body' that must be grouped" [p. 85].

Leonard's chapter on organization would still be valuable reading for curriculum planners and textbook writers of the 1980s.

It may seem odd that such a radical book concludes with a chapter on errors, which Leonard calls "Minimal Essentials." Yet it soon becomes apparent that Leonard's rethinking of this topic is as fresh and new as the rest of the book. He distinguishes immediately between two types of error:

1. What we may by correction brand as socially acceptable or unacceptable forms—positively right or wrong.
2. What we may better criticize as simply more or less clear or forcible and pleasing ways of expression.

Leonard's is one of the first pedagogical approaches to embody a sophisticated understanding of levels of usage, as well as to assume that mistakes in word choice are not somehow wrong in themselves. Classroom emphasis must be on issues that society deems unacceptable, and here only 100 percent accuracy will do. Other issues of style (in which Leonard seems distinctly less interested) can be dealt with in conference. Encouraged by research showing that a remarkably high percentage of errors were due to the misuse of a very small number of words, Leonard calls vigorously for similar research on those few key errors; he hopes to identify the major ones in order that English courses will be developed "in the light of broader knowledge of what is acceptable English usage to-day" (p. 120). As it turned out, the most important studies of error and usage done in Leonard's lifetime were those he conducted himself.

Because a large number of mistakes come from a relatively few forms, Leonard advocates concentrating on a few crucial forms a year, eradicating the errors by making their accurate use a matter of habit. This entails having teachers pass over in silence a number of minor errors in order to concentrate on the few under attack, another reason to end "the regime of wholesale red ink." Leonard wants a realistic standard; for it to be enforceable, it has to be both minimal and attainable both by drill and by use in the sentence. For instance, he wants students to be able to master W. F. Jones's standard list of the "one hundred demons of the English language" by grade six.[8] Punctuation drill would concentrate on sentence boundaries and, at the same time, focus on the class's judgment of effective expression. (Leonard considers most fragments and run-ons to be punctuation errors, not "sentence structure" problems as handbooks have it.) He recommends sentence combining (which he calls "sentence-massing," p. 158) as a means of building syntactic fluency, but he also cautions that long, complex sentences are not necessarily better than simple ones.

Again and again Leonard connects writing assignments to meaning and grammar to students' needs. He cites the psychologists' warning that "calling attention to a form before we are ready to give full, unremitting attention to its establishment can probably have no other result than confusion and repression" (p. 136). Leonard is caustic on the subject of vocabulary lists, for they require from the student "a rather robust idea of what he wants to say" before they are of use. Leonard illustrates his point by citing a teacher who notices an isolated example of the word *dwell* in a student paper and then encourages the rest of the class to use it in order to provide a nice variation on a familiar term. He remarks that matters might have been worse, "the teacher might have hit upon *abide*, which is rather less possible to find real use for." Leonard calls this whole approach "a wasteful misdirection of . . . energies" and asks (p. 169) "does the word *dwell* represent a possibly valid distinction, or add anything, except a pretty archaic flavor, to . . . expression? And . . . isn't it often the scorn with which healthy children regard this sort of thing that makes most of them relapse into content with fourth-grade vocabularies, even for writing university freshman themes?" For Leonard, language facts must meet the "test of use"; they must connect to what students need in order to express their ideas.

English Composition as a Social Problem is a humane, radical book. Though there is no sign that James Britton and other British composition theorists ever read Leonard, he anticipates much of their approach and practice. And it is instructive to compare his volume to an influential British book of its time, George Sampson's *English for the English* (1921), an example of a thoroughly sensible reform approach to composition. Sampson's viewpoint is a good bit more patrician than Leonard's; he places himself in the Arnoldian tradition and bears some trace of a missionary's impulse to lead the uneducated to a world of finer thinking and better expression. Leonard, on the other hand, is more radical, more reliant on the social setting to encourage learning, and much more willing to limit instruction to elements of language that students need in their present lives. All the chapter epigraphs in *English Composition as a Social Problem* are from Emerson, reminding us of how thoroughly Leonard's book is imbued with an American outlook.

Leonard was a prolific essayist in the decade 1916–1926, with articles appearing very regularly, especially in *The English Journal*. These were usually about grammar and usage and most often attacked some misguided notions of teaching or the upholding of outdated notions of propriety in language. For instance, the 1918 *English Journal* essay "Old Purist Junk" shows that expressions condemned by current textbooks are approved even by the most conservative of contemporary dictionaries. "More About Usage" (*English Journal*, 1918) is a review essay urging more research on what contemporary usage actually is. Leonard's essays and his stance on language and teaching gave him a reputation both as a follower of Dewey's social approach to composition and an articulate spokesperson for teaching students a modern, relativist view of usage. In 1918, R. L. Lyman of the University of Chicago characterized Leonard as the leading representative of the "social motives" school that believes that fluency and accuracy in expression "must be expected to come largely as byproducts of social situations. . . ."[9]

Leonard published his second book on the English curriculum in 1922, soon after his appointment as assistant professor at Wisconsin. *Essential Principles of Teaching Reading and Literature* is a compendium of Leonard's ideas on grade school and high school English instruction. He explicitly refers to Dewey (and lists him often in the notes and appendixes) and to Fred Newton Scott of Carpenter, Baker, and Scott, authors of the then-standard work on teaching English. In this book, Leonard combines a Deweyan outlook with an impressive grasp of the latest scholarship. Dewey's influence appears in the stress on contemporary readings; in the large role given to student activities, especially drama; in the refusal to make the study of literature a formal discipline; and in the continuing emphasis on the students' experiences as the basis for instruction as well as the goal of the whole English curriculum. And the scholarship appears throughout, not just added on to impress. Every page seems to contain a footnote to contemporary research on literature, reading, or learning theory; a whole chapter is devoted to a sophisticated discussion of reading tests.

According to Leonard, literature and reading instruction should be conducted in orthodox Deweyan fashion, starting with children's actual experiences and interests. A teacher must know what children actually

like even though children's tastes do not necessarily run toward the
works of value. For acquainting students with literature, Leonard, like
Dewey before him and Moffett after him, stresses plenty of group work
and active engagement, including having students perform drama and
compose their own poems and stories. In dealing with reading selections, he urges teachers as far as possible to "let the writer himself do
his own introducing" (p. 259). The instructor should leave out background material unless it is absolutely necessary, and even then the
introduction and background matter must themselves be real experiences, not just lists of historical dates or facts. This emphasis goes
entirely counter to the tendency at the time to spend pages on "information" before getting to the text at issue and to base examination
questions on that background information ("What was Milton's ninth
poem?" went a famous example) instead of on the work itself.[10]

Leonard's book also provides extensive reading lists for students. In
his suggested readings he freely mixes traditional works with items
drawn from science and technology. (He laments the dearth of interesting books about science for young people.) In poetry he recommends
the modern Americans Edward Arlington Robinson and Robert Frost
as poets to teach to young people. The inclusion of Frost is another
example of Leonard's up-to-date thinking; Frost was growing in popu-
larity in 1922, but he was far from being current among schoolteachers
and was certainly not taught in many university literature courses.

Essential Principles is impressive for another reason as well, its treatment of readings appropriate for boys and for girls. To be sure, it
contains some of the sexism that suffused educational approaches of its
time; there is an assumption that boys and girls will inevitably like
different books, for instance, and Leonard seems to be much more
confident about boys' tastes (even though his only child was a daughter). But the book speaks out strongly against the naive stereotyping—
the heroic boy, the emotional girl—then popular, making its sexism
pale in comparison with books of its time. And Leonard derides books
that display heroic qualities solely in the midst of armed combat or in
outlandish adventure stories. Here he shows signs of the pacifist streak
that animates his *Poems of the War and the Peace* (1921), a remarkable
collection of World War I verse.[11] Leonard's attitudes toward sex stereotyping and naive heroics, like his approach to reading and literature, demonstrate his forward looking sensibilies.

The second stage of Leonard's career, his scholarship on language
usage, seems at first to mark a major shift in direction. Some of his
writings between 1915 and 1926 had dealt with issues of correctness,
but mainly in the context of teaching. After 1927, though, his significant publications on teaching stop, and his work on language becomes

much more scholarly. The reasons for this shift must be a matter for speculation, but it is not hard to guess at some causes. First, Leonard may have said what he wanted to say about teaching. *English Composition as a Social Problem* and *Essential Principles of Teaching Reading and Literature* were clear statements of an educational approach. Without an abrupt change of philosophy, it is hard to see what more Leonard could have added, except some filling in of the details. Another reason for the shift may well have been Leonard's public position in the NCTE, which must have impressed on him the vital need for an indisputable standard of language instruction in a nation with burgeoning school enrollments and loud cries for efficiency. Unless informed people led the movement for better instruction, he might have felt, change would take the direction of more traditionalism and rigidity. Another motive might have come from the stress on empirical research during the 1920s. He had always displayed respect for researchers, filling his books with references to the latest studies. He may have wanted to participate in what looked like an outburst of "scientific" studies. Finally, the changing atmosphere in the Wisconsin English Department might have influenced Leonard to take on more obviously "scholarly" research because his colleagues were rapidly moving away from composition and toward purely literary or linguistic studies. As an indication of this dramatic shift, in 1909 everyone in the Wisconsin English Department taught freshman composition; in 1929, thanks to an influx of teaching assistants, only one senior professor taught it.[12]

But one should not make too much of the shift in directions, for the sharpened focus on language and usage grew directly out of an interest Leonard had had all along. And all Leonard's language studies seem ultimately aimed at the composition instructor who has to determine just what "language facts" to explain to the class. In that sense, Leonard's career moves in a clear and unified direction toward a deeper understanding of everything needed to teach composition effectively.

Leonard's research into language use grew directly from the conflict over what kind of English usage to teach in schools. Enrollments had been rising dramatically for a generation; educators had to train large numbers of new teachers, and the perennial issues of grammar and usage sharpened into the new question of what English was most appropriate for twentieth century students. Recent work in linguistics, particularly studies by George Philip Krapp, was moving away from reliance on traditional authorities and toward basing grammar and usage decisions on how educated people used the language.[13] This new line of thinking forced the issue: Should modern educators choose usages hallowed by time and upheld by authorities, or should they focus teaching on the language actually employed by the educated public? No friend of entrenched authority, Leonard long ago had joined the fight for a modern standard of English usage and for a "scientific" means of determining what usage actually existed.

By choosing to campaign for an up-to-date standard of usage, Leonard had joined the movement for modernizing the school curriculum. This movement, which went by the unfortunate name of "efficiency," placed heavy emphasis on making all school subjects useful, sometimes in a valuable, refreshing way and sometimes in a thoughtless, anti-intellectual manner.[14] In English instruction, the positive side of the efficiency movement led to ridding the curriculum of unnecessary material and leaving only what were called "minimum essentials" of grammar and usage; it also entailed a stronger relationship between the reading and composition curriculum and student interests. Widespread enthusiasm for the efficiency movement influenced the NCTE in 1921 to form its Commission on Minimum Essentials of English Teaching, partly perhaps to ward off criticism of school programs and partly no doubt to gain control of a powerful and potentially destructive trend. Leonard, by then a well-known advocate of removing inessentials, chaired this committee for most of the 1920s.

For Leonard, the issue of minimum standards depended on finding out what actual usage was. He had stated over and over that textbooks were full of outdated advice on usage, "junk" that should be removed wholesale. The task for researchers was to determine what was essential and what was junk, precisely the thing research studies promised to decide. The first major study on "essentials" had been conducted by W. W. Charters in Kansas City in 1915; it was soon followed by a host of others, some done by leading educators of the time like R. L. Lyman of Chicago and M. V. O'Shea of Wisconsin.

Unfortunately, most usage studies of the 1920s were based on a completely wrongheaded understanding of language, an outlook as old-fashioned as the textbooks Leonard attacked so bitterly. Thus, Stormzand and O'Shea's *How Much English Grammar* (1924), a typical study, employed a list of errors that would have satisfied the most hidebound traditionalist. Another error study concluded that incorrect use of hyphens accounted for $4\frac{1}{2}$ percent of all errors; and *shall* and *will,* for 2 percent of all grammar errors.[15] Its conclusion was that because *shall* and *will* appeared so often as errors, they deserved extra drilling. To someone like Leonard, who determined acceptability by educated usage, this must have seemed nonsense; the very fact that *shall* and *will* were confused so often showed that the distinction between them was disappearing, because when everyone makes a "mistake," it means that the usage in question has become the standard. But the studies of the 1920s relied on a static notion of language and completely lacked a sense of how usage changes over time and of which errors matter most to readers.

In their extensive 1928 review of research on errors, Leonard and

his coauthor, Dudley Miles, welcome the emphasis on removing inessentials but add drily that "practically all the studies of 'errors' need to
be done over in the light of recent linguistic inquiries into what really
are errors."[16] Ironically, then, the most up-to-date research was serving
to uphold the most outdated notions of error by employing such archaic categories. Leonard's own subsequent work on usage would respond to this situation by undercutting traditional error studies in a
devastating manner. By the end of the decade, he was to demonstrate
beyond doubt that many of the words and usages that researchers
considered to be errors were in constant use among educated Americans. Leonard would move the issue away from which errors students
made most often and toward the deeper and more interesting question
of how and why errors get labeled.

Leonard began his research program by immersing himself in historical studies; he went to Columbia to get his Ph.D. with George Philip
Krapp, then the most prominent American scholar on the history of
the language and the author of *Modern English* (1909) and *The English
Language in America* (1925), the standard, groundbreaking works in the
field. For his dissertation, Leonard sought to discover the source of
current notions of correctness and undertook a lengthy and comprehensive investigation into the grammars of the past. His resulting dissertation, *The Doctrine of Correctness in English Usage* (1929), published in
the monograph series "Wisconsin Studies in Language and Literature,"
is still the fullest exploration of the subject.

The Doctrine is a comprehensive history of how late eighteenth century grammarians codified English and imposed the rigid standards of
correctness that have persisted to the present. Leonard discovered that
though earlier centuries had witnessed scattered attention to grammar,
relatively few mentions of "correct" grammar or "correct usage" occurred until the eighteenth century. And though Swift and others
spoke about correct English in the early eighteenth century, it was the
latter half of the century that saw a "flood of English grammars," over
200 titles (p. 12). Leonard based his history on a close examination of
90 of these works, which ranged from brief articles and pamphlets to
Lord Monboddo's six-volume study.

From Leonard's perspective, eighteenth century grammars were
written without an understanding of how usage operates. Most were by
"clergymen, retired gentlemen, and amateur philosophers . . . [who]
had little or no conception of the history and relations of the classical or
other languages, and of course no equipment for carrying on linguistic
research or even for making valid observations of contemporary usage"
(p. 14). As a result, these grammarians relied on "unscientific" procedures and "ipse dixit" pronouncements. The book is a careful traversal
of their whole body of work.

Leonard notes from the beginning that almost all eighteenth century grammarians paid lip service to Horace's famous dictum in the *Ars Poetica* that usage, not authority, must rule speech. Campbell, for instance, in *The Philosophy of Rhetoric* (1776), followed Horace in developing what seemed like reasonable criteria for accepting change in language, arguing that grammars should be based on "national, reputable, and present use." But Leonard shows how Campbell applied his three criteria so narrowly that, instead of allowing contemporary usage to determine correctness, his criteria, applied the way he recommended, "resulted in a complete repudiation of usage" (p. 165). Similarly, Leonard describes prodigious efforts to multiply differentiae, for example, "to keep separate the various parts of speech and the preterites and past participles of verbs. This attempt was based in a belief that multiplication of forms is always an aid to expression; and no doubt it had its ultimate root in the reverence for the inflected classical languages" (p. 59). In practice, this meant such absurdities as introducing the dative case into English simply because it existed in Latin.

In *Doctrine of Correctness*, Leonard is particularly scornful of dogmatic attempts to enforce individual opinions about usage. Of the grammarian Robert Baker, he writes, "Most often he is authoritatively dogmatic, the true intellectual ancestor of the makers of purist handbooks" (p. 41). And in the same vein he points to example after example of "ipse dixit" pronouncements. Most notable was the phenomenon of writing "corrected" editions of poems, the best-known being Bentley's seventeenth century "correction" of *Paradise Lost*, which had its eighteenth century counterparts in the "remarks" on the style of authors by grammarians eager to supply examples of nice distinctions between words and of elegant usages. Thus, Hugh Blair's *Lectures on Rhetoric and Belles Lettres* (1783) devotes four lectures, some eighty-five pages, to a minute critique of the style of *The Spectator* and of Swift's *A Modest Proposal*.

Leonard seems especially delighted to demonstrate how grammarians failed to follow their own logic. He admits there might have been some "useful betterment" of the language if grammarians had followed their rule of analogy, which stated that like terms should be handled in a similar manner. Thus, "we might now have fewer cases of divided and anomalous forms such as those of 'ring' [ring, rang, rung; ring, ringed, ringed] and 'lie' [lie, lay, laid; lie, lay, lain; lie, lied, lied]" (p. 236). Ultimately, though, Leonard discerns little of value in the codification that occurred in the century, and he notes that the work of the grammarians did not lead to many visible changes in the language itself. His example is the prose of Dryden and Addison, written before the grammarians' assaults, but in most senses indistinguishable from modern English. In an appendix, Leonard lists over 300 prescriptions on grammatical and logical issues and concludes that of the 300, "less

than a dozen of those condemned constructions are actually regarded
as illiterate or popular usage today. In fact, a better point-score than
that of these rhetoricians and grammarians could certainly have been
achieved by pure chance" (p. 237).

Throughout *Doctrine of Correctness* runs an alternative approach to
determining standards of usage, the one provided in the seventeenth
century by John Locke. Leonard argues that "in the work of Locke and
his followers in eighteenth-century philosophy [i.e., Berkeley and
Hume] . . . appeared clear statements that language is solely 'a form of
behavior' originated by compact and determined in form and meaning
by convention. Here was a thoroughly adequate basis for the destruc-
tion of theological and mystic notions and for beginning a sound and
scientific study of language problems" (p. 26).

The one eighteenth century grammar that actually followed Locke's
lead was not by a grammarian but by a renowned chemist, Joseph
Priestley, who published his *Course of Lectures on Oratory and Criticism* in
1777. Here was someone who possessed the scientific temperament
for the job of analyzing usage as it existed and for criticizing the
shoddy methods employed by the amateur grammarians. Leonard
states, "Priestley is undoubtedly the first writer in English, and apparently the only one in the eighteenth century, to take a clear and
reasonably consistent view of usage. His grammar alone makes a beginning at carrying out in a sensible and practical way the philosophy
of Locke . . ." (p. 142). Leonard cites with approval Priestley's criticism
of his contemporaries: "Our grammarians appear to me to have acted
precipitately in this business of writing a grammar of the language.
This will never be effected by the arbitrary rules of any man, or body
of men whatever" (p. 144). And though Leonard disagrees with some
of Priestley's points, he states that

> he will no doubt be counted right in his view as to what the intelligent
> student of language should do in the meantime. Priestley's scientific train-
> ing stood him in good stead here, and an examination of his judgments in
> detail confirms the belief that he for the most part actually carried out his
> promise of proposing conjectures on the trend of the data he examined,
> rather than uttering dogmatic rules [pp. 144–145].

For Leonard, the difference between the authoritarian outlook of
the vast majority of eighteenth century grammarians and the scientific
approach taken by Locke and Priestly represents a clash of two views
of grammar and, ultimately, of two different views of the relationship
between mind and the world. Drawing on Ogden and Richards' treatment of Locke in *The Meaning of Meaning*, Leonard maintains that
language is a convention and that the relationship between word and
idea, or between signifier and signified in modern terms, is totally

arbitrary. For Leonard as for Locke, appropriate usage is determined by custom; a grammarian could understand what good usage actually is simply by collecting enough facts. Thus, Leonard's view of language studies is inductive: One collects enough data and thus proves one's point. (This is exactly what *Current English Usage* was to attempt.) Opposed to this stands the traditional view, traceable to St. Augustine's *On Christian Doctrine*, which was espoused by so many in the eighteenth century: Language is an entity that reflects the outside world, with the correspondence between signified and signifier divinely ordained. The study of such an entity should proceed rationally, not empirically. And the methodology should be deductive, not inductive. Indeed, it is possible to make a list of the oppositions Leonard suggests or actually includes:

Traditional	*Modern*
rational	empirical
deductive	inductive
authority	custom
hierarchical	democratic
ancient	modern
rule	practice

Needless to say, Leonard's own bent was for the terms on the right.

Doctrine of Correctness contains the richest statement of Leonard's beliefs about the role of language research. If eighteenth century grammarians had followed Locke, he writes, they would have shifted language studies toward an examination of barriers to real communication. Leonard finds in Locke a powerful antidote to the narrow grammatical studies, an alternative approach that could "occupy the highest capacities of mind." Leonard sums up Locke's legacy:

> Effectiveness of communication is not an affair of logical precision, extreme differentiation of words or inflections, or consistency in etymology. It depends chiefly on defining terms and illustrating them fully, detecting fallacies hatched by wrong uses of words as of logic, and using expressions only with intent and consistent heed to their meaning [p. 243].

And Locke's legacy points toward a modern approach,

> the outline of a fundamental grammar such as Ogden and Richards present. Their techniques of translating and expanding symbols constitute a rigorous discipline. . . . For obviously, learning something of the mastery of a living and indefinitely adaptable medium, and of its adjustment to the thought and passions of men in actual communication, calls for a more difficult sort of mental activity than following precise and invariable logical prescriptions [pp. 243–244].

Despite its powerful analysis of how grammarians consistently chose rigidity and authority, *Doctrine of Correctness* seems curiously incomplete. Missing is an analysis of *why* so many rigid grammar texts appeared when they did. Leonard recognizes the validity of the question; he had said at the outset, "[I]t seems necessary to find out if possible why such prescriptions became prevalent and popular, and especially, upon what assumptions about language, and the English language in particular, they are based" (p. 12). But he never delves into the former question; instead, he devotes the whole book to explaining eighteenth century assumptions toward the English language. In short, he never connects this outburst of grammar with social changes occurring at the time: the rise of popular education and increased literacy rates, the increased opportunity for upward mobility, the expansion of the middle classes. These were popular topics for historical analysis in the 1920s, but Leonard decided to keep his focus on the narrow linguistic level.

Nonetheless, Leonard provides the basis for such an analysis, though the material appears scattered throughout the book. For instance, Leonard supplies plenty of evidence that better grammar and usage were class issues. He cites Philip Withers writing in 1788: "Purity and Politness of Expression . . . is the only external Distinction which remains between a Gentleman and a Valet; a Lady and a Mantua-maker" (p. 169). And one of the hacks who wrote grammatical treatises, Reverend John Trusler (*Distinctions Between Words Esteemed Synonomous in the English Language,* 1783), had been the "diligent compiler" of *The Way to Be Rich and Respectable* and *Principles of Politeness.* Most grammatical works were self-help books for people on the way up who did not want to seem "low" or provincial. Key words were "easy, proper, elegant." And there was a serious assault on provincialisms as well, especially Scotticisms (p. 178).[17]

Also missing is some sense of the economics of grammar. For instance, it is evident that these authors were responding to a a very special marketplace. Schoolmasters, often poorly educated themselves, needed authoritative textbooks for growing numbers of students; they wanted prescriptive canons of usage, not the democratic and relativistic notions of custom that Leonard approved of. Schools were in the business of raising their charges' standing in the world, not leveling the standards of society as a whole. Grammar in the eighteenth century was a product. One thinks of Mencken's treatment of Noah Webster as grammarian in *The American Language* and realizes how much more Leonard could have supplied. Yet one suspects that more complex historical questions did not really interest Leonard very much. At bottom he was a activist teacher in twentieth century America, faced with students whose immediate needs were far removed from those of eighteenth century England.

The inductive approach championed in *Doctrine of Correctness* makes its fullest appearance in Leonard's last and most famous book, *Current English Usage* (1932).[18] We may guess at Leonard's reasons for writing it from a passage at the end of *Doctrine of Correctness,* in which he tells how eighteenth century attitudes toward correctness persisted:

> The nineteenth century saw a yet completer inundation, with the stimulus of Kantian categories and of half-digested ideas out of Indo-Germanic philology. And with all this addition, whole heaps of the prejudices, taboos, and prescriptions to eighteenth-century writers were carried entire into the books of writers who followed them, so that a majority of their ideas, based as most of these were in metaphysical and illogical concepts of language and usage, may nevertheless be found today earnestly and convincedly taught in schools. A cleaning out of this ancient purist muddle is suggested as essential before we can do any effectual teaching of composition [p. 238].[19]

Leonard's task in *Current English Usage* is to begin that cleaning out.[20] His book is a compilation of informed opinion about the acceptability of many disputed items in punctuation (Part I) and in grammar (Part II). For punctuation, he listed 81 items, "chiefly . . . periods, capitals, commas, and semicolons" (p. 3) and collected judgments from 144 editors of books, magazines, and newspapers. For each item, Leonard provided a sentence and asked the judges to state whether the particular punctuation mark highlighted was "required," "preferred," "permitted," or "not allowed." For grammar and usage, Leonard listed 230 disputed items and collected judgments from 229 editors, authors, linguists, school and college English teachers, and businesspeople.[21] Judges were given sample sentences and asked to decide which of four categories a highlighted term fitted: literary English; standard colloquial English; trade or technical English; or popular, uncultivated English. They were explicitly directed *not* to state what a term's category *should* be. Thus, there was a significant difference between the two parts: Part I, though it asked working journalists and editors to describe their actual practice, still permitted a frank expression of preference, but Part II aimed at being purely descriptive.[22]

Current English Usage is the first large-scale survey of informed opinion about what punctuation, grammar, and usage people actually employed. It is a remarkable-looking book, fairly bristling with charts, diagrams and statistical tables. Leonard reproduces the precise wording of the questionnaires, along with the names of the respondants and a full tabulation of individual responses. There is an air of scientific objectivity about the enterprise, much like the articles to be found in today's *Research in the Teaching of English.* But, of course, a survey of expert opinion about English usage is indeed reducible to charts and tables. And the statistics, graphs, and full methodological explanations serve another purpose, replicability. It becomes a straightforward matter to check Leonard's procedures and see how he derived his conclusions. The effort to discover what usage actually is seems one place where such scientific procedures are called for, particularly because, as Leonard himself had demonstrated, the whole subject of usage had been so clouded over with misinformation and ipse-dixit pronouncements of ignorant authorities.

The punctuation section relies on professional editors and journalists, experts for whom matters of pointing are everyday affairs. Why it appears first is open to conjecture, but it is worth noting that error studies of the 1920s consistently found that over half of student errors were in punctuation, with placement of commas most to blame. If one regards Leonard's attempt to record current usage as a challenge to contemporary textbooks and research studies, punctuation deserves to come first because it looms so large in texts and error counts.

But, as always with Leonard, issues of correctness and punctuation connect readily with questions of meaning. As he puts it in *Current English Usage,*

> The problem of communication may be put in the form of a question: "How can I be sure that my reader as he reads will think of the same things I am thinking as I write?" It takes at least two to make a meaning; a meaning has not been established until the reader has understood. [Punctuation] is one of the main devices at the writer's command for clear representation of thought. . . . Meaning comes first; pointing must adapt itself to meaning [pp. 76, 78].[23]

Leonard's discoveries about punctuation come as no surprise. Professionals were far more willing to approve so-called "disputed" items, especially lack of commas after introductory phrases and between closely related independent clauses. Leonard's major conclusion is entirely in keeping with his desire to simplify: "[T]he first clear, practical conclusion that can be drawn from the ballots . . . is that there are comparatively few demands for 'required' pointing. Seldom are more than half the votes under the heading of 'required'" (p. 72). Other votes were in the categories of "preferred" or "permitted," but very few were for "not allowed." In other words, experts were much freer in their practice than teachers and textbooks. That finding causes Leonard to stress the need for a thorough rethinking of the grammatical definitions of the sentence and to call for new definitions that would "establish a mind-set toward the dynamic application of thinking to writing . . ." (p. 83). And at the end of the chapter, Leonard returns to an old theme, the relationship of correctness to the readers' expectations that he had detailed in *English Composition as a Social Problem* in 1917: "[P]erhaps the clearest implication of this study is that the way to learn to punctuate is to write with a real audience in mind and to test the effect of what has been written on that audience" (p. 85). The punctuation chapter demonstrates conclusively that the strict rules promulgated by the handbooks were far out of line with the practice of those who determined the style of the popular and literary press. And it suggests new, provocative ideas about refining educators' and linguists' understandings of grammatical definitions.

Grammar and usage are inherently more complex subjects than punctuation and much less susceptible to technical answers. Accordingly, Leonard employed a wider group of judges, including English teachers, authors, linguists, and businesspeople. Panelists were directed to score "according to your observation of what is actual usage rather than your opinion of what usage should be." They were asked to rate words in sample passages according to four levels of usage:

1. Formally correct English, appropriate chiefly for serious and important occasions, whether in speech or writing; usually called "Literary English."
2. Fully acceptable English for informal conversation, correspondence, and all other writing of well-bred ease; not wholly appropriate for occasions of literary dignity: "standard, cultivated colloquial English."
3. Commercial, foreign, scientific, or other technical uses, limited in area of comprehensibility; not used outside their particular area by cultivated speakers: "trade or technical English."
4. Popular or illiterate speech, not used by persons who wish to pass as cultivated, save to represent uneducated speech, or to be jocose; here taken to include slang or argot, and dialect forms not admissible to the standard or cultivated area; usually called "vulgar English," but with no implication necessarily of the current meaning of vulgar: "naif, popular, or uncultivated English" [p. 187].

As it turned out, few examples of "trade or technical English" appeared, so the decisions were really among the three: formal, cultivated colloquial, and illiterate.

The inclusion of cultivated colloquial English as an approved category marks a departure. All error studies of the 1920s had made simple yes-no distinctions; a term either was or was not an error. But Leonard, conscious of the varying levels of usage employed in America, set up his categories in order to allow for good informal as well as good formal English. It is worth remembering that "colloquial" in the 1920s was often a term of opprobrium; *Webster's New International Dictionary* (known familiarly as Webster's Second), when it appeared in 1932, would use "colloquial" as a usage label for terms that were not fully acceptable. By inventing a category called "standard, cultivated colloquial English" and blandly stating that it was "fully acceptable English for informal conversation, correspondence, and all other writing of well-bred ease," Leonard was following his long-held conviction that a vigorous colloquialism was more suitable than formal, literary English. Nonetheless, his description of the colloquial begs a lot of questions. Just what is "informal conversation," for instance? Isn't all conversation, by its very nature, informal? And for traditionalists, the phrase "standard, cultivated colloquial" was a contradiction in terms; nothing could be "standard" and "colloquial" at the same time. Leonard's category represents the viewpoint of linguists of his time, and it accurately captures twentieth century practice. But we should be aware that his decision to create the "colloquial" category reflects an ideological stance that was in favor of a richer, fuller notion of permissible styles.

The results in the grammar and usage section fully supported Leonard's contention that English teachers were sorely out-of-date. Time and again the majority of his judges determined that usages scorned by textbooks were fully established as "cultivated colloquial," including "It is *me*," "Invite *whoever* you like to the party," and "Everyone was there, but *they* all went home early." Similarly, his judges determined that established usage indiscriminately employed *will* for *shall* and *can* for *may*, among other sturdy examples of textbook errors.

Leonard's findings are dramatic evidence that textbooks for school and college work make far greater demands than professional editors, linguists, and businesspeople. And he draws the inevitable practical conclusions:

> ... usages on which the judges strongly agree can be profitably taught; in regard to usages upon which the judges are evenly divided, dogmatism is unjustified. Extensive drill on either form of a divided usage would clearly be a loss of time; and it is equally obvious that no class time should henceforth be wasted in an effort to eradicate any construction here listed as established—no matter what the personal preference of the instructor or the dictum of the adopted text [p. 187].

Current English Usage's concluding statement linking grammar to science is perhaps addressed to teachers and textbook authors:

> Laws of grammar should be like physical laws ... they are what observation has determined to be the case; they are descriptive. If closer observation shows the behavior is different from what was formerly thought, the scientist stands ready to alter the law. The same is true of grammar; if people change the way they talk or write, then the law changes. And some of the usages in this book that have been judged acceptable are thus no longer violations; we have to change the law [p. 189].

This seems in keeping with the inductive, "scientific" spirit Leonard admired in Locke and Priestley. It is precisely the approach that upholders of traditional distinctions despise.[24] However, the passage seems unsubtle and ultimately unsatisfying, for it leaves out what *Doctrine of Correctness* makes clear, the power of old forms over the mind and the attractions associated with authority. And it also neglects the question of who enforces the law. Social communities are notoriously complex, much more protean than scientific communities. The ground rules in both are different, as are the stakes. A comparison between scientific laws and laws of usage needs to include the complexities not just of laws but also of enforcement. One has to wonder if the passage was written by Leonard himself or by one of the book's compilers.

Current English Usage began a trend that lasted for a decade. Leonard's search for the facts of usage was continued by Marckwardt and Wolcott, whose NCTE book *Facts About Current English Usage* (1935) was explicitly designed as an updating of Leonard's work. In 1933, Leonard's successor at Wisconsin, Robert Pooley, published *Grammar and Usage in Textbooks on English*, a critical look at how much outdated information was passed along in English texts. And the examination of usage culminated with Charles Carpenter Fries's superb *American English Usage* (1940), the most famous and influential study of how Americans actually wrote and spoke.[25] The search for an understanding of contemporary American usage had been begun by Krapp at Columbia early in the century. Leonard published the first book-length empirical study of the subject, and Fries supplied the crowning touch, the definitive book. It is easy to surmise that Leonard's contribution might have been greater had he lived, but what he did was achievement enough.

In his forty-three years, Leonard accomplished a great deal, enough to fill a longer life span. His pedagogy alone would rank him as an important figure in composition, as would his research on language. With the two put together Leonard becomes unique, someone who accomplished work of the first rank in both pedagogy and research. As a man of his time, he sums up some of the best tendencies of the 1920s; and as a figure in the history of composition, he speaks to us still.

Notes

1. A brief but useful biography by a friend and colleague of Leonard at Wisconsin, William Ellery Leonard, appears in Dumas Malone, ed., *Dictionary of American Biography* (New York: Scribner's, 1933), vol. 11, pp. 178–179. Hereafter it is referred to as WEL.
2. Leonard's colleague Robert Pooley claims that Leonard was "perhaps the most important American correspondent of Ogden and Richards before Richards' coming to the United States." Pooley's sketch of Leonard appears in *Word Study*, 28:2 (December 1952), pp. 1–3.
3. William Ellery Leonard's account of the accident appears in WEL, p. 179.

Richards described the drowning in a letter to the Madison *Capital Times* of 17 July 1931.

4. Dewey, whose difficulties with writing were notorious, devoted little space to the subject, important though he acknowledged it to be.

5. For Leonard, as for Moffett half a century later, argument papers that verge on debates are sterile exercises aimed at winning points rather than on truly changing opinions of a audience.

6. Like almost all writers who rely on the students' own knowledge and experience to generate composition, Leonard omits any detailed treatment of invention. This is in keeping with his Deweyan attitude, which assumes that students invariably have plenty of things to say, but need proper encouragement from teachers and peers as well as help in shaping their statements to meet the audience's needs.

7. Leonard also anticipates Jerome Bruner's spiral curriculum of the 1960s when he writes of making "development a slowly upward spiral movement, in which we return again and again to the same sorts of problems in story-telling or explanation or discussion and meeting of common difficulties—but each time discover, together with more complex difficulties, greater power and sureness in handling them and in judging our own and other people's results" (p. 64).

8. In keeping with his interest in modernizing the curriculum, Leonard continually advocates a vigorous colloquialism as opposed to a dull, literary formal English. This was to have significant consequences later in his career, when he argued that "standard, cultivated colloquial English" was just as acceptable as formal, "literary" English. See page ooo for further discussion.

9. "Fluency, Accuracy, and General Excellence in English Composition," *The School Review*, 26 (January 1918), p. 87. Lyman was to become NCTE president in 1931.

10. Leonard's high school drama anthology, *The Atlantic Book of Modern Plays* (Boston, 1921), carries out the approach of *Essential Principles*. His introductions are short, and his questions for discussion (pp. 281–283) brief but very helpful.

11. *Poems of the War and the Peace* (New York: Harcourt Brace, 1921). In his introduction, Leonard writes that young people must know war "not as a glorious and jolly adventure, but as the sternest and most terrible reality in human history" (xiii).

12. Warner Taylor, "A National Survey of Conditions in Freshman English," *University of Wisconsin Bureau of Educational Research Bulletin*, 2 (May 1929), p. 19.

13. Edward Finegan's *Attitudes Toward English Usage* (New York: Teachers College Press, 1980) provides a fine overview of changing attitudes toward usage and an excellent bibliography. For similar views of the usage controversies of the 1930s, see Albert Marckwardt, "The Professional Organization and the School Language Program (NCTE)," *Linguistics: An International Review* (April 1977), pp. 107–123, and Raven I. McDavid, ed., *An Examination of the Attitudes of the NCTE Toward Language*, NCTE Research Report No. 4 (Champaign: NCTE, 1965), especially Chapters 1 and 2.

14. "Efficiency" seems to have been a catchword throughout the decade 1915–1925, much as "process" was in the 1970s. The best treatment of the progressives' attempts to modernize the curriculum during the 1920s is Lawrence Cremin's masterful *The Transformation of the School* (New York: Knopf, 1961). See also Raymond E. Callahan, *Education and the Cult of Efficiency* (Chicago: University of Chicago Press, 1962).

15. Macon Anderson Lieper, *A Diagnostic Study of the Errors Made by College Freshmen in Their Written Compositions,* George Peabody College for Teachers Contributions to Education, No. 22 (Nashville, 1926).

16. Dudley Miles and Sterling Leonard, "Research in High School English," Chapter 16 of *Sixth Yearbook, Department of Superintendence* (Washington, D.C.: National Education Association, 1928), p. 327. Miles followed Leonard as NCTE president in 1927.

17. Here Webster, the American, was the exception; he practically reveled in differences between British and American English, though as Leonard says, "His determination to erect a national custom largely out of New England materials was laudable, but of a piece with his zeal in reforming pronunciations and word-choices" (p. 180).

18. *Current English Usage* was initiated by Leonard but completed after his death by a committee of the NCTE. The book bears his name on the title page, for he conceived of, designed, and began carrying out the study; it is very much his book. But it is not always easy to tell exactly which portions were actually written by him. And it is impossible to guess whether the final text would have been altered had he lived to finish it. The acknowledgments on pp. xxi–xxii detail the help of a great many collaborators.

19. The dissertation by Leonard's student Mildred Hergenhan, "The Doctrine of Correctness in English Usage in the Nineteenth Century" (Wisconsin, 1938), traces how eighteenth century doctrines survived until the early twentieth century.

20. In fact, Leonard had begun the task in 1927 with "Current Definitions of Levels in English Usage," *English Journal,* 16 (1927), pp. 345–359, a collaboration with H. Y. Moffett that employed the same methods as *Current English Usage.*

21. There were two grammar ballots. The first, on the 102 items Leonard and Moffett had listed in their 1927 article, was answered by 229 judges. The second, on 130 additional items, was answered by 17 linguists and 32 NCTE members who had also responded to the first ballot.

22. At times some judges in Part II, after giving opinions on what was acceptable, could not resist adding their own preferences (p. 153).

23. Echoes of Ogden and Richards' *The Meaning of Meaning* pervade the conclusion. For Leonard's very early ideas on this subject, before he read Ogden and Richards, see "The Rationale of Punctuation: A Criticism," *Educational Review,* 51 (January 1916), p. 89.

24. See Jacques Barzun, *The House of Intellect* (New York: Harper, 1959), in which he attacks the most famous of Leonard's successors in language research, Charles Carpenter Fries, as "the theorist who engineered the demise of grammar in the American schools" (p. 241).

25. Excellent summaries of this decade's work appear in Finegan and in McDavid. Fries headed the College Section of the NCTE in 1926; his plans for the book are mentioned in accounts of the NCTE conventions of 1925 and 1926. See *English Journal* 15 (1926), pp. 69–74, and 16 (1927), p. 64.

Selected Publications of Sterling Andrus Leonard

BOOKS

English Composition as a Social Problem. Boston: Houghton Mifflin, 1917.
> Leonard's most thorough treatment of composition pedagogy, it contains chapters on basing composition on students' activities and experience, on relying on group work in the classroom, on organizing ideas, and on correct grammar and usage. A radical book, perhaps the best translation of John Dewey's educational ideas into composition pedagogy.

The Atlantic Book of Modern Plays. Boston: Little, Brown, 1921. Reprint: Philadelphia: R. West, 1979.
> An anthology aimed at student audiences, it contains short plays by Galsworthy, Lady Gregory, Synge, Yeats, and O'Neill, among others.

Poems of the War and Peace. New York: Harcourt, Brace, 1921.
> Includes poems by Alfred Noyes, E. V. Lucas, Wilfred Owen, Siegfried Sassoon, Rupert Brooke, Joyce Kilmer, Robert Graves, Edith Wharton, and Robert Frost, among others. Notable for its pacifist tone and superb selection. Leonard acknowledges help from Carl Van Doren, William Ellery Leonard, and Louis Untermeyer.

Essential Principles of Teaching Reading and Literature, Philadelphia: J. B. Lippincott, 1922.
> A compendious (460-page) teaching guide aimed at intermediate and high school teachers. Leonard places heavy emphasis on drama and on students' experiences. Includes a chapter analyzing reading tests. Concludes with 100 pages of bibliography.

General Language (with Riah Fagan Cox). Chicago and New York: Rand McNally, 1925.
> A text on grammar, word study, and etymology, including a brief history of the English language. Intended primarily for junior high school students.

The Doctrine of Correctness in English Usage, 1700–1800. University of Wisconsin Studies in Language and Literature, No. 25. Madison: University of Wisconsin Press, 1929. Reprint: New York: Russell and Russell, 1962.
> Leonard's Columbia dissertation, done under the direction of George Philip Krapp. A 360-page account of how eighteenth century England witnessed the rise of uniform notions of grammatical correctness. A scholarly, heavily documented attack on the narrowness and dogmatism of most eighteenth century grammarians.

Current English Usage. National Council of Teachers of English: English Monograph No.1. Chicago: Inland Press, 1932.
> Completed after Leonard's death by an NCTE committee. Two major parts: (1) Punctuation and (2) Grammar and Usage. The first book-

length survey of the subject. Based on the responses of experts to questionnaires about the current state of English usage. Leonard found that many terms and usages condemned by textbooks were well established.

Leonard collaborated on some other textbooks and wrote introductions for a number of editions and anthologies.

SELECTED ARTICLES

"Old Purist Junk." *English Journal,* 7 (1918), pp. 295–302.

> Attack on outmoded information in texts. Leonard found that words the textbook authors warned against were cited without disparagement in the *Oxford English Dictionary.*

"More About Usage." *English Journal* 7 (1918), pp. 481–484.

> Review essay on R. P. Utter's *Everyday Words and Their Uses* (1916) and J. Leslie Hall's *English Usage* (1917).

"Building a Scale of Purely Composition Quality." *English Journal* 14 (1925), pp. 760–775.

> Leonard's first attempt to devise a meaningful method of measuring quality in speech and writing. He urged testers and teachers to treat form [i.e., errors] and "composition proper" "at different times and with different procedures."

"The Wisconsin Tests of Sentence Recognition." *English Journal* 15 (1926), pp. 348–357; "The Wisconsin Tests of Grammatical Correctness." *English Journal* 15 (1926), pp. 430–442.

> The two Wisconsin Tests were developed to deal with major issues of correctness: sentence fragments and common grammatical problems. They represent a departure from the common methods of testing and are worth renewed interest.

"English Teaching Faces the Future." *English Journal* 16 (1927), pp. 2–9.

> Leonard's NCTE presidential address to the Philadelphia convention in November 1926. A capsule summary of Leonard's pedagogical attitudes at the time he was moving toward linguistic scholarship. Characteristically, he draws on Aristotle, not for rules but to to demonstrate that, as the *Rhetoric* makes clear, "composition occurs not in a vacuum but in a social situation."

"Current Definitions of Levels in English Usage" (with H. Y. Moffett). *English Journal* 16 (1927), pp. 345–359.

> The first appearance of the research methodology that would be incorporated in *Current English Usage.*

"Research in High School English" (with Dudley Miles). Chapter 16 of *Sixth Yearbook, Department of Superintendence.* Washington, D.C.: National Education Association, 1928.

> A thorough review of research efforts done on reading, writing, and literary studies at the high school level.

KENNETH BURKE

by William F. Irmscher

In a century of increasing specialization, Kenneth Burke can scarcely be labeled. He has been a short story writer, novelist, poet, essayist, translator, literary critic, art critic, music critic, social critic, historian, teacher—and the list might well be extended. Yet in Burke's own terms, *rhetorician* subsumes all of these roles, for the rhetorician transcends diversity in the search for unity.[1] R. P. Blackmur says of Burke: "I think he would rather be called Rhetor, as honorific and as descriptive, than anything else."[2]

Burke has observed that the growing emphasis on literary criticism in the early part of the century caused rhetoric to be parceled out among various disciplines such as anthropology, social psychology, sociology, psychoanalysis, and semantics. He calls them the "new sciences." Burke has attempted single-handedly—single-mindedly—to restore the maimed art of rhetoric by studying the other disciplines to bring together again a new rhetoric "reinvigorated by fresh insights which the 'new sciences' contributed to the subject."[3] I. A. Richards and Burke are the two figures who again give repute to the word *rhetoric* in the twentieth century. Burke was strongly influenced by Richards in the role of critic and rhetorician, but that relationship need not be traced because, in characteristic fashion, Burke found his own course. In so doing, he invented his own field of study. Stanley Edgar Hyman remarks, "The reason reviewers and editors have had such trouble fastening on Burke's field is that he has no field, unless it be Burkology."[4]

It would be more accurate to say that Burke has extended the range of rhetoric to make it a study of all human relations, he has drawn architectonic dimensions for rhetoric to make it a world view for those who conceive of it narrowly, and he has transformed the key term of rhetoric from "persuasion" to "identification."[5] In the preface to *Rhetoric and Writing* (1965), W. Ross Winterowd writes, "When I discovered Kenneth Burke, the parts of my accumulated knowledge of my subject fell into place, for it is the genius of Burke that he has shown how rhetoric is a controlling function in human activity."[6] This kind of personal testimonial suggests the synthesis Burke accomplished both for himself and others.

Burke's predilection for syncretism may have grown out of a personal need to pull together his own largely unsystematic reading. The breadth of his reading is, to say the least, intimidating. He once wrote, "All is grist to the mill, so use as your appetite prompts."[7] One can only speculate to what degree Burke's lack of formal university training fostered his individual style of thinking and writing. He spent one semester at Ohio State in 1916 and a year at Columbia in 1916–17, where he received encouragement from John Erskine. He was, therefore, not limited by anyone's prescribed reading list, nor was he stamped with the educational philosophy of a particular university. In fact, Burke's discontent with the university stemmed from the fact that he was required to take certain courses when he wanted to take others—one, like medieval Latin, which was part of the graduate program, and another in the *Greek Anthology* that presumably was not offered.

Burke finally bargained with his father to give him a portion of the money his father was paying for fees at Columbia, and he would get a place in Greenwich Village and keep up with his studies. He joined several former high school buddies who rented in the Village and began the pursuit of an education that transcended the limitations of the typical university's approach to the liberal arts in terms of distribution requirements. His own studies ultimately gave him the broad perspectives and insights of men like Aristotle, who served as the prototype for his pursuit of knowledge; Coleridge, whom he studied and absorbed; Spengler, who he says scared him and impressed him greatly;[8] Marx, from whom he pragmatically borrowed; and Freud, who represented one of the major influences on his thinking.

If Burke had not been freed from the usual restraints and rigors of the academic system, he might well have been less adventurous, less intrepid—the very intrepidness that made him vulnerable to attack by precise scholars, who could scorn his lax use of texts, and by traditional readers, who sought logical connections in his prose, missing the emerging vision of the whole. In *A Rhetoric of Motives,* Burke describes his method:

> So we must keep trying anything and everything, improvising, borrowing from others, developing from others, dialectically using one text as comment upon another, schematizing; using the incentive to new wander-

ings, returning from these excursions to schematize again, being oversub-
tle when the straining seems to promise some further glimpse, and making
amends by reduction to very simple anecdotes.[9]

The description prompts Kermit Lansner to comment: "A strange com-
bination of system and no system."[10] It was left to academic advocates
like Austin Warren, Stanley Edgar Hyman, and Marie Hochmuch Nich-
ols to recognize that Burke's thinking was producing nothing less than
a comprehensive view of human affairs.

Apart from his own voracious appetite for reading from an early age,
Burke derived encouragement from a group of friends, including Malcolm Cowley, William Carlos Williams, and Hart Crane, some of whom
became lifelong associates. Malcolm Cowley calls Burke his "oldest
friend."[11] The two of them met when Cowley was three years old and
Burke was four. Cowley's father was the Burkes' family physician. Malcolm and Kenneth attended Peabody High School in East Liberty, Pittsburgh, where they and several other young men represented "the literary crowd," who, as Cowley describes the coterie, made good grades in
English, read unassigned books, and helped edit the school magazine.
They considered themselves different from other boys, dreamed of
adventures in romantic places, and wrote poems. They were already
becoming the artists they destined themselves to be.[12]

After high school, the Cowley-Burke connection continued. Cowley
went to Harvard, but the two continued to correspond and to see one
another when circumstances permitted. After Burke's brief stint with
college, he continued to read and write on his own. At one period when
he stayed with his parents in Weehawken, New Jersey, Cowley describes Burke's daily schedule: "Kenneth wrote in the mornings—
stories, poems, essays, fables, plays, all of them lopsided, brilliant, im-
mature, and full of characters who explained themselves in paradoxes."
In the afternoons, he studied and wrote letters.[13] Burke planned his life
so that he could eventually go to France to breathe the air that had
nourished Remy de Gourmont, Flaubert, other French novelists, and
the French symbolists, all of whom he read and admired. He was immersed in European culture. He interspersed his letters with French
and German phrases. He discovered Thomas Mann, a major influence
on his own fiction, and Arthur Schnitzler, both of whom he translated
during the early 1920s.

Unfortunately, Burke never fulfilled his dream of getting to Europe
at that time, even by means of military service in World War I. He was
rejected for service, worked in a shipyard, and continued to write. Even
though Burke had published poetry in the literary magazine at Ohio
State in 1916 and has continued to write poetry throughout his life, his
early literary successes began about 1919 with the publication of short

stories in a number of little magazines: *Smart Set, Broom, The Little Review, Secession,* and *S₄N.*

Most important, however, was the acceptance of "Mrs. Maecenas" by *The Dial,* a story that had been rejected by the editors of *Smart Set.* "I later found good cause to be grateful for its rejection," Burke writes, "since its acceptance by *The Dial* marked the entrance of that magazine into my life—and for me that was almost as momentous a moment as the act, or accident, of being born."[14] Because *The Dial* was the outstanding literary review of its time, Burke's comment is probably not overstated. His writing earned him a place on the staff of that magazine. In the years that followed, he quite literally pursued his education in public by reviewing an incredible range of books that extended far beyond the limits of imaginative literature. But of special importance was Burke's absorption in everything he read. William Rueckert, one of the most discerning scholars of Burke's writing, says, "He almost never reviewed a book he did not in some way use."[15] Burke continued in various capacities with *The Dial*—as reviewer, music and art critic, translator, and editor—until 1929, when the magazine ceased publication because of the economic crisis. During the decade of the 1920s, however, it kept Burke intimately in touch with a broad segment of the intellectual life of New York City.

The climax of Burke's association with *The Dial* came when, at the urging of Marianne Moore, the editor from 1926–29, Burke was given the Dial Award for distinguished services to American letters, formerly given to figures like Sherwood Anderson, Van Wyck Brooks, T. S. Eliot, Ezra Pound, William Carlos Williams, and Marianne Moore herself. The extent to which Marianne Moore believed in Burke is described in words by William Wasserstrom:

> In her view, he was the one American critic who possessed technique and sensibility enough to reconcile art and culture, science and imagination in a single theory of literary value which is simultaneously a theory of human virtue in a comprehensive but probably not final sense of that bloody word.[16]

In that comprehensiveness also lay Burke's potential as a rhetorician.

Burke's proximity to New York kept him in touch with Malcolm Cowley and other young writers and intellectuals, who met informally on occasion to talk, read to one another, and carouse. Burke later refers to the association as "a pranksome kind of literary 'gang morality' loosely linking several young writers who liked to think of themselves as a monster-loving 'advance guard.' "[17] The group, however, was for him an antidote to his early intense individualism. Burke abandoned what he calls his "Aestheticist period" and began to develop new interests in social communication and political issues.[18]

Burke often refers to the stock market crash of 1929 as traumatic and the years of the Great Depression in the early 1930s as times of great stress, even though, he adds, he never missed a meal. But the conditions he observed established his anticapitalism and invited his involvement with Communism. In 1935, Burke delivered a lecture entitled "Revolutionary Symbolism in America" to the First American Writers Congress in New York, an address that antagonized both capitalists and Communists alike. Burke's brand of Communism represented the humanitarian idealism characteristic of the philosophical position of many intellectuals in the 1930s. When Burke realized the consequences of the Stalinist purges in the Soviet Union, however, he changed his position. When *Permanence and Change,* originally published in 1935, was revised in 1954, Burke deleted the passage proposing a congregational state, an ideal community based on material cooperation. But the consequences of Burke's previous commitment could not be easily erased, and the Communist tinge continued to plague him during the McCarthy era in the post-World War II days.

From 1943 on, when Burke joined the faculty of Bennington College, Vermont, he more and more divorced himself from political affiliations although no reader of Burke's major works can ignore his continuing strong interest in politics as an important arena in which rhetoric functions. Bennington introduced Burke to a new circle of friends. Colleagues like Howard Nemerov, Stanley Edgar Hyman, and Francis Fergusson were especially important to Burke because all of them wrote sympathetic critiques of his work that were largely responsible for drawing national attention to the uniqueness of his thought.

Understanding critics were necessary because Burke's style was—and always has been—an annoyance to many readers. Sidney Hook, one of Burke's severest critics, begins his review of *Attitudes Toward History* (1937) with the lines: "The greatest difficulty that confronts the reader of Burke is finding out what he means. His individual sentences seem to be clear, but when put together they are obscure, sometimes opaque."[19] Austin Warren touches on the troublesome element in Burke's style when he writes: "Certain themes fascinate this mind, and it plays with them, turns them about, drops them, returns to them. There are digressions and episodes—pensées."[20] Burke poses difficulty for at least two kinds of readers: (1) those like Sidney Hook, who demand linear, logically connected prose and who, therefore, are unable to involve themselves in Burke's performative style, a manner intended to engage the reader in the generation and presentation of ideas; and (2) those who are skimmers or, as Burke calls them, "evaders," who prefer the "impromptu" to the "studied," "a method without risk,"[21] and who accordingly find anything abstract and theoretical inaccessible because, again, they do not become adequately involved. The fact remains that most of what has been written about Burke is, with few exceptions, more difficult to read than Burke himself. His concepts do not lend themselves easily to paraphrase or abridgment. There is no substitute for reading Burke himself. Cowley says, "Burke is one of the authors who write to be read twice."[22] One reason is that the expansiveness with which he treats an idea is the source of understanding it. His style is part of the meaning.

Burke did not see literature and rhetoric as antithetical. He approached rhetoric through literature: rhetoric as a natural and consistent feature of literature. He writes, ". . . effective literature could be nothing else but rhetoric: thus the resistance to rhetoric *qua* rhetoric must be due to a faulty diagnosis."[23] By 1932, the date of the publication of Burke's novel *Towards a Better Life,* he had already published a book of short fiction, *The White Oxen* (1924), and his important collection of critical essays *Counter-Statement* (1931), a work of special interest to the rhetorician because of its discussion of form as the psychology of the audience. His definition of form is often quoted: "*Form* in literature is an arousing and fulfillment of desires. A work has form in so far as one part of it leads a reader to anticipate another part, to be gratified by the sequence."[24]

Even though Burke began as a writer of imaginative literature and turned early to literary criticism, Malcolm Cowley does not view him primarily as a literary critic,[25] and neither does R. P. Blackmur, who writes that Burke is "only a literary critic in unguarded moments, when rhetoric nods."[26] Saying that he could not accomplish his purpose in a realistic, objective work of fiction, Burke writes in the Preface to the first edition of *Towards a Better Life:* "So I reversed the process, emphasizing the essayistic rather than the narrative . . . the ceremonious, formalized, 'declarmatory.' "[27] Burke refers to the work as an antinovel. Thus, even as novelist, he was quintessentially the rhetorician.

Burke's poems written between 1915 and 1954, published in 1955 under the title *Book of Moments,* have a similar rhetorical connection.[28] Burke speaks of poems as "moments" because they are times that need to be fulfilled before he returns to other work. He writes, "Lyrics are 'moments' insofar as they pause to sum up a motive."[29] One is reminded of his description in *A Rhetoric of Motives* of the ineffable moment of pure persuasion—"where the pendulum is at rest, not hanging, but poised exactly above the fulcrum. It is the change of direction, from systole to diastole, made permanent . . . it is the pause at the window, before descending into the street."[30] We can assume that a poem was such a moment of pure persuasion.

Whether Burke was writing poems, stories, or critical essays, he was constantly writing himself into a synthesis that would become his system. Cowley calls the critical essays "seedful,"[31] and indeed dramatism,

the theory with which Burke is most closely associated, was the out-
growth of his own creative and critical experience.[32]

On almost any occasion, speaking or writing, when Burke has undertaken to explain his ideas, he has begun with the definition of the
human being. The most complete statement is contained in his essay
"Definition of Man," originally published in *The Hudson Review* (Winter
1963–64) and later made the lead essay in the collection *Language as
Symbolic Action* (1966). Burke explains at the beginning of the essay that
any definition "so sums things up that all the properties attributed to
the thing defined can be as though 'derived' from the definition. In
actual development, the definition may be the last thing a writer hits
upon. Or it may be formulated somewhere along the line. But logically
it is prior to the observations that it summarizes."[33] In this sense,
Burke's definition of man is the basis on which almost all else is built.
He writes the definition in four "clauses" and adds what he calls a "wry
codicil":

> Man is
> the symbol-using (symbol-making, symbol misusing) animal
> inventor of the negative (or moralized by the negative)
> separated from his natural condition by instruments of his own making
> goaded by the spirit of hierarchy (or moved by the sense of order)
> and rotten with perfection.[34]

The first two clauses establish the language-based center of Burke's
thinking. Only the symbol-using being as a creature of free will is
capable of symbolic action as opposed to the motiveless motion of other
creatures and inanimate objects in nature. Only the user of a symbol
system is capable of perceiving the negative dimension of anything—
that which is not—an abstraction. The capacity to perceive the negative
is a mark of humanness and a source of ethical behavior, as Burke
discusses in other works.

The third clause becomes the basis for Burke's position on issues
such as capitalism and technology—those conditions that corrupt the
essential substance of the human being. To illustrate, he views war as a
"*special case of peace*—not as a primary motive in itself, not as *essentially*
real, but purely as a *derivative* condition, a *perversion*."[35] Just as heat is
the absence of cold—it is not essentially real—so war is the absence of
peace. The human creature is peace-loving by nature. War disturbs
that essential nature. The purpose of *A Grammar of Motives* is stated in a
motto: *Ad Bellum Purificandum*.[36] It is a dedication of the book to the
cause of purifying war, as the alchemist purified base metals to produce
gold.

The fourth clause goes further to say that this same being is "goaded by the spirit of hierarchy." The human being is *homo dialecticus,* purposive. As a symbol-using animal, the creature is a "transcending animal."[37] Burke thus provides an Upward Way—his phrase—in the total scheme of things. His motto *Per linguam, praeter linguam* (By and through language, beyond language) points to the special role that the study of language assumes in the self-fulfillment of individuals and in their social cohesion.[38]

When Burke refers to the codicil "rotten with perfection" as wry, he colors the phrase with irony. The drive and compulsion of the human being to attain perfection often lead to disaster. Burke sees this tendency in society's current pursuit of technology without regard for the consequences to human life. He catches the ultimate irony of this mania in the last stanza of a modernized jingle:

> And if the great military man
> Took the great thermo-nuclear warhead
> And put it into the great intercontinental ballistic missile
> And dropped it on the great land mass,
> What great PROGRESS that would be![39]

On a solemn note, he writes in "Definition of Man":

> The best I can do is state my belief that things might be improved somewhat if enough people began thinking along the lines of this definition; my belief that, if such an approach could be perfected by many kinds of critics and educators and self-admonishers in general, things might be a little less ominous than otherwise.[40]

In personal correspondence, Burke summarizes his views on humanity: "I still hold to the three miracles: that there is existence at all, that our kind worked out the sexes as it did, and that we found our way with words."[41] This is Burke's faith in the symbol-using capacity of the human being. The approach, of course, is dialectical by means of language—a give-and-take that leads to a resolution of differences and the ultimate fulfillment of a state of perfection, however remote it may seem from present world chaos. In *A Rhetoric of Motives,* Burke writes, "Rhetoric is concerned with the state of Babel after the Fall."[42] We remain in that "state of Babel," but the purpose of rhetoric is through dialectic to restore the *lingua Adamica,* at which point rhetoric will have dispensed of its purpose for being.

Burke envisions the restoration of this prelapsarian state as an Upward Way. He conceives of his own works as a verbal pyramid, patterned after Aristotle, one side of which represents the logical, one the rhetorical, the third the poetical, and the fourth the ethical.[43] The base of the pyramid is Babel, the "Human Barnyard," as he calls it in *A Rhetoric of Motives*. The apex is the ultimate point of transcendence, the point of summation, oneness, and common understanding. The movement must be upward if the distance and division between the sides are to be narrowed. The Upward Way—the reach for order and peace—is the essential motive of the symbol-using animal. The pyramid is the comprehensive diagram of Burke's thought and writings.

A Grammar of Motives, the first of these major works, was published in 1947. Its purpose was to provide the terms for the study of human relations. As early as 1935, Burke had already determined the source of those terms. In *Permanence and Change,* he writes: "The conclusion we should draw from our thesis is a belief that the ultimate metaphor for discussing the universe and man's relations to it must be the poetic or dramatic metaphor."[44] In *A Grammar of Motives,* Burke sets forth his dramatistic approach by asking a simple question: "What is involved, when we say what people are doing and why they are doing it?" and then proceeds to identify the five key terms of dramatism: act, agent, scene, agency, and purpose. These terms have come to be known as Burke's pentad although Burke explains that they have precedent in Aristotle's four causes and in the questions of the medieval scholastics: *quis* (agent), *quid* (act), *ubi* (scene defined as place), *quibus auxiliis* (agency), *cur* (purpose), *quo modo* (manner, "attitude"), *quando* (scene defined temporally).[45] Burke describes them as "basic forms of thought" that provide a perspective for the analysis of any situation involving motive.

The terms are simple and self-explanatory: The human being is an *agent* capable of *action* in a social *setting* by varying *means* for diverse *purposes.* Burke allows that the agent may be impersonal—a force or institution or influence. If we focus on *act*—"foremost among equals"— in terms of all the other elements of drama, we may begin to understand human motives more completely.

The pentad is thus an analytical and heuristic device. Burke cautions that it can be used "profoundly or trivially," a reminder to those practitioners who choose to see in the pentad little more than an investigatory tool like the journalistic formula and overlook the potential of it as a logical method.

A Grammar of Motives is an extended elaboration of the implications of the terms, especially as the five terms yield the possibility of ten ratios (scene-act, scene-agent, scene-agency, scene-purpose, act-purpose, act-agent, act-agency, agent-purpose, agent-agency, and agency-purpose). The ratios give flexibility to the terms. Any broadening or narrowing of them—what Burke calls the scope or circumference— changes the perspective and the generative possibilities. Dramatism, Burke explains, "aims always to make us sensitive to the 'ideas' lurking

in 'things' which might even as social motives seem reducible to their
sheerly material nature, unless we can perfect techniques for disclosing
their 'enigmatic' or 'emblematic' dimension."[46] A confrontation between
two nations, for example, may be at first dismissed simply as a border
skirmish, but, if it is investigated in terms of the dramatistic scheme, it
may then be seen as a highly complex situation involving agents, co-
agents, and counteragents, indicating motives far more complex than
the initial explanation. One need only think of the complexity of the
Middle Eastern conflict that has drawn world attention, especially since
the establishment of Israel in 1948.

Conversely, what at other times seems infinitely confused and complex when first observed may become understandable if the five basic
terms are reclaimed in all their simplicity and applied to the situation.
Burke tells of seeing an aerial photograph of two launches included in
an exhibition at the Museum of Modern Art. The wakes of the vessels
on the water formed an infinity of lines, but, says Burke, "the picture
gave an impression of great simplicity, because one could quickly perceive the generating principle of the design."[47] He intends the pentad
to act as that kind of generating principle, to translate confusion to
understanding.

At various times, Burke has tried to explain how the terms should be
thought of. In a note to *The Rhetoric of Religion* (1961), he says that the
terms are to be thought of as questions. He adds, "They are really but a set
of *blanks to be filled out.* They are an algebra, not an arithmetic."[48] Burke
maintains that the blanks can be filled out differently to form various philosophies. Part II of the *Grammar* illustrates how the terms can be merged
or certain ones given logical priority to the exclusion of others in order to
shape different schools of thought. Thus, he demonstrates how dramatism represents a methodology and shapes world views.

The *Grammar* provides the logical grounding for the works to follow;
Burke promises a *Rhetoric* and a *Symbolic.* In 1950, *A Rhetoric of Motives*
appeared. To Burke, anything rhetorical implies a "you and me" quality, but he expands on the traditional notion of classical rhetoric as
language addressed to real audiences. To Burke, rhetoric also includes
self-persuasiveness for motives of one's own choosing, either deliberate
or subconscious. An audience can be oneself, the "I addressing the
me."[49] Thereby, Burke introduces a new psychological dimension to
rhetoric.

The key term of *A Rhetoric of Motives* is identification. Instead of the
kind of persuasion that attempts to change others by explicit design or
to manipulate, Burke envisions a more subtle means of drawing the
"you" and the "me" together by having them identify with one another
in terms of qualities or motives that they share in common. Whatever
our differences, we may share a common ethnic background or a place

of residence or a social group or a belief or future goal. Identification is thus a way of bridging the differences that separate us. In Burke's own terms, rhetoric is rooted in "the use of language as a symbolic means of inducing cooperation in beings that by nature respond to symbol."[50] Identification becomes a means of social cohesion and a step toward the ultimate goal of universal oneness.

A synonym for identification in the *Rhetoric* is consubstantiality, a term that builds on Burke's concept of "substance," defined in *A Grammar of Motives*. Substance is the essential "whatness" of any person or thing, not in explicit scientific terms, but in terms of what we believe characterizes or differentiates something. In terms of aim, for example, teacher and preacher might more readily identify with one another than poet and pedant. Burke admits that our notions of "substance" are often vague and ambiguous, but he sees that ambiguity as a resource of rhetoric, for consubstantiality—an acting-together of "substances"—can be accomplished even when the differences at first appear to be great.[51]

Identification or consubstantiality, which the apex of the verbal pyramid represents, leads Burke in Part III of the *Rhetoric* to discuss pure persuasion, which he says "involves the saying of something, not for an extra-verbal advantage to be got by the saying [recall the aim of traditional persuasion], but because of a satisfaction intrinsic to the saying. . . . For if union is complete," asks Burke, "what incentive can there be for appeal?"[52]

The dialectical process idealistically culminates in absolute communion. In more realistic terms, human beings participate in a relative kind of communion by sharing ultimate terms. In order to make clear their effect, Burke arranges words in a hierarchial order: first, positive terms, the names of visible and tangible things; second, dialectical terms, including generalizations and titular words like "Elizabethan" or "capitalism," terms that have no tangible referent; and, finally, ultimate terms, summarizing words or unitary principles that act as a generating force in the lives of people. Burke also calls them god-terms—words like *money* or *progress* or *technology*, which we deify and use to validate actions. If society accepts *progress* as a god-term, then to say something is progressive is to endorse it and give it credence. In Burke's words, god-terms interpret "all incidental things in terms of the over-all fulfillment towards which the entire development is said to be striving."[53] God-terms have rhetorical force.

Toward the end of the *Rhetoric*, Burke writes, "By going by the verbal route, from words for positive things to titles, then to an order among titles, and finally to the title of titles, we come as far as rhetoric-and-dialectic can take us, which is as far as the book contracts to take us."[54] Burke, however, was not through with his interest in ultimate terms. In 1961, he published *The Rhetoric of Religion*, subtitled "Studies in Logol-

ogy." He explains that logology is "studies in words-about-words."[55]
Logology is central; all other studies and specializations radiate from it.
Burke chooses theology as his topic for study because it is preeminently
"words about God"—God being the Word (*Logos*), whether or not the
word designates a being who exists or not.

Throughout his career, the Word has been Burke's deity. Burke has
been a servant of words. He has identified himself as a "word man."[56]
His ultimate faith has been in words. He confesses, ". . . I couldn't fold
up in the Church despite my great love of theology. . . ."[57] Yet he could
not resist the logological perfection of theology, especially Christian
theology. He expresses his credo in a "Dialectician's Hymn," a prayer in
the manner of "pure courtship, homage in general, the ultimate idea of
an audience, without thought of advantage, but sheerly through love of
the exercise"[58]:

> Hail to Thee, Logos,
> Thou Vast Almighty Title,
> In Whose name we conjure—
> Our acts the partial representatives
> Of Thy whole act.
>
> May we be Thy delegates
> In parliament assembled.
> Parts of Thy wholeness.
> And in our conflicts
> Correcting one another.
> By study of our errors
> Gaining Revelation.
>
> . . .
>
> May we compete with one another,
> To speak for Thy Creation with more justice—
> Cooperating in this competition
> Until our naming
> Gives voice correctly.
> And how things are
> And how we say things are
> Are one.
>
> . . .
>
> And may we have neither the mania of the One
> Nor the delirium of the Many—
> But both the Union and the Diversity—
> The Title and the manifold details that arise
> As that Title is restated
> In the narrative of History.
> Not forgetting that the Title represents the story's Sequence,
> And that the Sequence represents the Power entitled.

For us
Thy name a Great Synecdoche,
Thy works a Grand Tautology.[59]

The Rhetoric of Religion ends with a dramatic piece entitled "Epilogue: Prologue in Heaven," a dialogue between The Lord and Satan (the positive principle and the negative principle). This masque is one of the most appealing and clear statements of Burke's thought. Burke seems determined to follow the words he gives to the Lord: "I'd rather not be so complicated in our discussion, since I am in essence simple."[60] He forgoes elaborations and complexities and, as simply as possible, explains again the fundamental attributes of the human being that he had set forth in his "Definition of Man." In recent years, Burke has said that he is "Occamizing" himself, reducing his thought to a few basic principles. The "Epilogue" is one example.

By the time *The Rhetoric of Religion* had been written and published in 1961, Burke had also completed and published "A Dramatistic View of the Origins of the Language" in two issues of *The Quarterly Journal of Speech* in 1952 and 1953. In the absence of a volume of ethics that he planned to call *On Human Relations,* Burke designates this long article, later included in *Language as Symbolic Action,* as his treatment of the ethical use of language.[61]

The essay is not so much a projection of the good life in terms of particular principles of moral behavior. Rather, it traces the means by which we determine values in terms of linguistic behavior, specifically in our ability to use the negative. Burke writes, "Reason is the ability to use the negative *qua* negative, the moralistic equivalent being the ability to distinguish between right and wrong."[62] The negative invites a choice between *yes* and *no.* Ethics is language-based by the choice we make between the two. Conscience translates *yes* to *right* and *no* to *wrong.*[63]

Burke builds his argument on the premise that "the negative must have begun as a rhetorical or hortatory function, *as with the negatives of the Ten Commandments.*"[64] The command "Thou shalt not" to creatures of free will, of course, invites the possibility of a disobedient no. Disobedience leads to guilt, which, in turn, sets into motion the ritual drama of mortification and victimhood in the human being's effort to alleviate guilt and seek redemption. Burke writes, "Logologically, the distinction between natural innocence and fallen man hinges about this problem of language and the negative. Eliminate language from nature, and there can be no moral disobedience."[65] Given the nature of language, Burke offers a compromise to the human being's inclination to choose only between polar opposites: ". . . where two opposed principles are being considered, each of which has the 'defects of its qualities,' what we want is something that avoids the typical vices of either and combines the typical virtues of both. Or, dialectical resources being what they are, we can readily propose that any troublesome *either-or* be transformed into a *both-and*."[66]

In the series of Motivorum books Burke proposed to publish, the third volume seems to have been bypassed. Although Burke never published the *Symbolic* in book form, it was planned in several versions and exists in scattered publications. The appendix to William Rueckert's *Kenneth Burke and the Drama of Human Relations* (2d ed.) includes three versions: one including essays Burke wrote from 1950–60, a second that includes additions up to 1967, and a third altering the contents further as of 1978. Furthermore, while teaching a summer course at the Indiana University School of Letters during the late 1950s, Burke distributed to his students a 391-page typescript of a text entitled "Poetics Dramatistically Considered." Instead of completing the *Symbolic,* Burke worked on collections like *Language as Symbolic Action* (1966) and *Philosophy of Literary Form* (1967), a new edition of a book originally published in 1941. Some of the selections in a list of thirty-four essays that Rueckert lists for the *Symbolic* are included in these volumes, but twenty-one out of the thirty-four remain uncollected.[67]

Thus, despite the unfinished form of the grand pyramidal structure, Burke did give thought to all its parameters, and by the mid-1960s it existed as a comprehensive scheme to be responded to, acted on, or ignored. Unfortunately, *The Rhetoric of Religion* was virtually ignored. The revival of interest in Burke that has accompanied the republication of his works by the University of California Press at Berkeley since the mid-1960s has continued to grow until Burke, perhaps for the first time in his long life, is being given the acclaim he deserves as one of the monumental thinkers of this century.[68] If we acknowledge one definition of rhetoric as a study of paradigms—the organizing principles of human relations; the implicit but inviolable rules of living in social harmony; the assumptions on which human communication, understanding, and identification depend—then Burke assumes a special role as the master rhetorician of this age.

If we think of the range of rhetoric in terms of a continuum with rhetorical theory at one extreme and applied rhetoric at the other, we may get some notion of the gulf that separates the two. Rhetorical theory, as Burke has dealt with it in his works, borders on philosophy. At times, it is philosophy. Applied rhetoric, on the other hand, which includes anything from electioneering to advertising, borders on education. In fact, teaching composition in the schools and colleges has become one of the most common forms of applied rhetoric.

What is Burke's connection with the teaching of composition in the American college?

Composition in the American college, at least as the large enterprise we know today, began in the aftermath of World War II. It posed vast administrative problems for the colleges. In time, it also invited serious deliberation of writing as a specialized discipline. These needs were addressed by the founding of the Conference on College Composition and Communication. Its first meeting took place in Chicago in 1950. Kenneth Burke was a featured speaker. He addressed the group on the topic "Rhetoric—Old and New."[69] In essence, he delivered the message of *A Rhetoric of Motives,* published also in 1950. One would like to think of this speech as a kind of proclamation to an organization that could have become the forum for Burkean thought throughout the 1950s and 1960s just as it became the forum for the new grammars. Yet Burke still remembers feeling "quite dismal about its bad reception."[70] There is no evidence that anyone saw a relation between what Burke was saying in theoretical terms and what they were doing in the day-to-day business of teaching writing. Burke's ideas were still remote from the concerns of the practitioners.

Yet, in a book entitled *Roots for a New Rhetoric,* published in 1959, Father Daniel Fogarty, a professor of education at St. Mary's University, Halifax, Canada, examined in some detail the theories of Richards, Burke, and a group of general semanticists, including Korzybski, Hayakawa, Wendell Johnson, and Irving Lee, to see what choices their ideas held for the classroom. Fogarty wanted to look ahead ten years to determine the adequacy of these theories for the first-year course in college composition and their capacity to meet the needs that "the great revolution in communication" was bringing about.[71] To Fogarty, new needs meant an expanded, more generalized course that would include philosophical concerns like abstraction, logic, definition, and epistemology, as well as concentration on syntax, style, and mechanics, usually associated with writing courses.

Fogarty, of course, knew of Burke's 1950 speech to the Conference on College Composition and Communication (CCCC) and recognized in it an extension of the concerns of rhetoric that would address his own. In the Foreword to the book, Francis Shoemaker, a professor at Teachers College, Columbia, characterizes CCCC at that time as an organization "seeking a scholarly rationale broad enough to embrace oral and written composition, approaches to literary criticism, media studies, and descriptive linguistics."[72] Fogarty's chapter on Burke, therefore, was intended to show in what way Burke's theories might provide such a scholarly rationale. In addition, Fogarty's book served as an introduction to Burke for many teachers. And Fogarty indicates in a footnote that Burke took "an enthusiastic interest" in the chapter and devoted time both to talk to Fogarty and to answer letters clarifying his positions.[73]

Fogarty summarizes his findings in a "Whole-view Chart of Rhetoric Possibilities and Choices," in which the first column lists the concerns of Aristotle's rhetoric, next the concerns of traditional composition courses of the 1950s, and finally the additions and changes that would result if the theories of Richards, Burke, or the General Semanticists were applied. (See Figure 1.) Fogarty says of Burke's aims: ". . . he protests that he does not want to do away with the old discipline, and that his philosophical elements are also expected to pervade the whole of the teaching rhetoric he envisions, because his 'identification' has taken over as central term and central aim in place of Aristotle's 'persuasion.' "[74] A statement of this kind indicates Burke's recognition of the need for a philosophy as the basis for an applied rhetoric, but Burke, of course, does not offer specific proposals as to how this is to be accomplished.

Toward the end of his book, Fogarty writes, "The cultural needs of the nineteen-sixties will probably determine the shape of the prose expression courses in colleges."[75] There is a certain irony in the innocence of Fogarty's words because the late 1950s gave no indication of the cultural upheaval that would mark the next decade. The ferment of the 1960s, however, provided the kind of situation that Burke's thought could address although it is doubtful that he was widely read. The concept of the interconnectedness of things—the broad circumference of human affairs—was pertinent, especially to university students who were not willing to think of the scholarly pursuits of the university as divorced from the harsh realities of Vietnam. They read special significance into words such as these from *A Rhetoric of Motives:*

> Hence, however "pure" one's motives may be actually, the impurities of identification lurking about the edges of such situations introduce a typical Rhetorical wrangle of the sort that can never be settled once and for all, but belongs in the field of moral controversy where men properly seek to "prove opposites."[76]

A passage of this kind suggested that the universities with their research and financial investments were an integral part of the scene, not removed from it. Burke's strong social consciousness, his pacifism, his anticapitalism, his antiscientism, his views of cooperative competition— these were all popular positions in the late 1960s and early 1970s. Obviously, Burke was not on any best-seller list, but those who were introduced to his works in college courses found a ready identity with the things he was saying. His thinking seemed right at a time when young people in particular were seeking new answers to old questions and concerning themselves with the future of society and humanity.

William Rueckert writes, "Burke's influence is widespread and essentially beneficial but very hard to trace. . . . Perhaps one cannot do more than acknowledge . . . some of the ways in which Burke has entered into the life of the mind in his own time."[77] This comment may be particularly true of Burke's influence on the teaching of composition in the American college. Actually, it may not even be possible to speak of "influence," but only to describe those conditions that prompted practical rhetoricians to look to Burke for whatever resources he might provide for an emerging discipline.

A new attitude toward composition can be dated from approximately 1965. A whole rethinking of traditional concepts began: the way we write paragraphs; the way we think of the entire process of writing, not just the product alone; the concept of writing as a way of learning and developing; the relation of aim and style; the ways in which thinking, speaking, and writing differ; how we think about error; the way we conceive of usage—all fundamental matters that have increased our understanding of how people generate thought, write, and appraise what they write. Professionals were not necessarily seeking a "scholarly rationale" broad enough to cover all the things English teachers do in English classes, as Francis Shoemaker seemed to think in the Foreword to *Roots for a New Rhetoric,* but they did begin to seek a theoretical base for the act of writing—a paradigm that would embrace and make meaningful a diversity of approaches to the act of writing and the teaching of composition.

Fogarty had envisioned a first-year composition course beginning with a philosophy of rhetoric and filtering down to the level of practice. Yet what usually occurs is just the opposite—a movement from practice to theory if theory is dealt with at all. The teaching of composition, we have learned, ought to begin in the act of writing, not in the study of grammar and usage or the study of rhetoric. Practice, however, need not be thought of as atheoretical. It may be based on unspoken, intuitive assumptions. The importance of formal theory is that it brings those assumptions to a conscious level of awareness so that they can be examined, even tested. Systematic theory following practice can bring understanding to the act of writing, just as philosophy lends meaning to living or logic gives shape to thinking. If the words on a page written by a freshman student seem incredibly remote from the theories of transcendence proposed in Burke, we can perhaps think of the Upward Way he speaks of. We can introduce students to different concepts of motive for writing. We can help them in their efforts to win identification through writing. We can enhance their views through Burke's perspectives by incongruity. We can direct their attention to themselves as an inner audience. We can emphasize dialogue, basic to dialectic, as a way of learning. We can demonstrate the generative force of the pentad as a way of knowing. And we can show that form is a means of appeal by which the writer can lead and ultimately satisfy the reader.

	I	II
	ARISTOTLE	CURRENT TRADITIONAL

PHILOSOPHY OF RHETORIC:

THOUGHT-WORD-
  THING RELATIONSHIPS
ABSTRACTION
DEFINITION
LOGIC
DIALECTIC

TEACHING RHETORIC:

The "Art" of Rhetoric	GRAMMAR
	SYNTAX
BOOK I	SPELLING
	PUNCTUATION
Rhetoric	MECHANICS
Kinds	
Persuasion	Four Modes of Discourse
Deliberative	Exposition
Forensic	Description
Epideictic	Argumentation
Deliberative	Narration
Orator's Needs	Argumentative
Good	Clarity
Happiness	Logic
Knowledge	Coherence
Epideictic	Style Qualities
Praise and Blame	Clearness
Virtues	Force
Amplification	Coherence
Forensic	Interest
Law and Wrong	Naturalness
Causes of Human Acts	Other Devices
Justice	Communication
Other Means	Symbols
	Word Counts
BOOK II	Concreteness
	Psychology of Communication
Emotion	Audience Reaction, etc.
The Fourteen Character Types	Divisions
Environments	Words
Topoi	Sentences
Modes of Persuasion	Paragraphs
Example	The Whole
Enthememe	Specialized Forms
Maxim	Letter
Concrete Topoi	Essay
Refutation	Speech
	News
BOOK III	Feature
	Advertising
Style	T.V. and Radio
Clear	Novel
Natural	Short Story, etc.
Appropriate	
Language	
Qualities	
Metaphor	
Simile	
Connotation	
Ambiguity	
Cadence, etc.	
Arrangement	
Divisions, etc.	

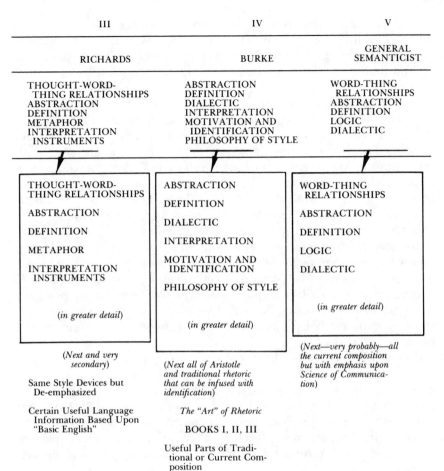

III	IV	V
RICHARDS	BURKE	GENERAL SEMANTICIST
THOUGHT-WORD-THING RELATIONSHIPS ABSTRACTION DEFINITION METAPHOR INTERPRETATION INSTRUMENTS	ABSTRACTION DEFINITION DIALECTIC INTERPRETATION MOTIVATION AND IDENTIFICATION PHILOSOPHY OF STYLE	WORD-THING RELATIONSHIPS ABSTRACTION DEFINITION LOGIC DIALECTIC
THOUGHT-WORD-THING RELATIONSHIPS ABSTRACTION DEFINITION METAPHOR INTERPRETATION INSTRUMENTS *(in greater detail)*	ABSTRACTION DEFINITION DIALECTIC INTERPRETATION MOTIVATION AND IDENTIFICATION PHILOSOPHY OF STYLE *(in greater detail)*	WORD-THING RELATIONSHIPS ABSTRACTION DEFINITION LOGIC DIALECTIC *(in greater detail)*
(Next and very secondary) Same Style Devices but De-emphasized Certain Useful Language Information Based Upon "Basic English"	*(Next all of Aristotle and traditional rhetoric that can be infused with identification)* The "Art" of Rhetoric BOOKS I, II, III Useful Parts of Traditional or Current Composition	*(Next—very probably—all the current composition but with emphasis upon Science of Communication)*

Figure 1. A whole-view chart of rhetoric possibilities and choices, also illustrating how the newer theories shift philosophies of rhetoric to teaching rhetoric. From Daniel Fogarty, *Roots for A New Rhetoric* (New York: Russell & Russell, 1959).

Furthermore, dramatism provides a model for the teaching of writing. Significantly, each begins with the act. In his essay on "Terministic Screens," Burke reminds us that our choice of terms for any particular act will define the limits of our perception of it. In contrasting dramatism and scientism, for instance, he writes, "Either approach ends by encroaching upon the territories claimed by the other. But the *way* is different, Dramatism beginning with problems of act, or form, and Scientism beginning with problems of knowledge, or perfection."[78] So, in order to test the implications of a definition, we might begin with this premise: Writing is a form of complex behavior, the complexity arising because it is an act in a particular setting, by an individual with a motive, by selected strategies that affect the outcome. If this premise is true, then the pentad becomes a paradigm for composition, and the terms in various ratios apply. Composition studies within the last twenty years have increasingly recognized the effect of the ratios on the act, especially writing in terms of agent-act (the effects of persona and voice on prose), scene-act (the relation of context and situation to the style of writing), purpose-act (the importance of motive as a starting point for writing), and agency-act (the recognition of invention as a necessary way of getting ready to write and continuing to generate thoughts from beginning to end).

Composition specialists have not consistently given thoughtful consideration to their choice of terms in speaking of writing. Burke, however, reminds us in "Terministic Screens" that "any nomenclature directs the attention into some channels rather than others."[79] Particular terms set up screens; they carry designated implications. Undoubtedly, much of the diversity in the teaching of composition may be explained in terms of the differences determined by a basic definition of the act:

> Writing is a skill.
> Writing is self-expression.
> Writing is creative activity.
> Writing is communication.
> Writing is a mode of thinking.
> Writing is a way of learning.
> Writing is a way of knowing.
> Writing is a form of complex behavior.

Clearly, writing is—or can be—all of these, but the implications of each are different for purposes of teaching. The difference between writing as a skill and writing as complex behavior is, in Burke's terms, a difference in circumference; one is narrower in scope than the other. But there is also a qualitative difference between an act that is virtually the equivalent of mechanical motion and an act that involves psychological, social, and rhetorical dimensions. Composition pedagogy, constantly concerned with usefulness, has given little recognition to the implication of the basic definition for the act itself.

The lack of a clear definition has also affected the kind of texts that have been used in the teaching of writing. Authors have borrowed ideas from various sources to shape eclectic texts, neglecting to express the rationale on which the selection has been made. Eclectic texts, of course, are safe because they are usable; they are marketable because publishers have learned that a book committed too singly to one philosophy or methodology limits its audience. One is reminded of Burke's cynical commentary in *A Rhetoric of Religion:* "In the search for academic preferment or for quick sales in the book mart, their [the Earth-People's] teachers and writers will slap together various oversimplified schemes that reduce human motives to a few drives or urges or itches involving food, sex, power, prestige and the like. . . ."[80] The consequence of "slapping together" has been to separate parts from wholes, to divorce practice from theory. Burke has always attempted to hold the two together and always to see specialized activities in broader contexts.

It is one thing to envision a new rhetoric, as Richards did in *The Philosophy of Rhetoric,* or to announce a new rhetoric, as Burke did in his speech to the first meeting of CCCC; it is another thing to find an audience receptive to what is new. Richards' ideas may have had more ready reception among teachers because the pedagogical implications were clear. Richards actually prepared materials for teachers in the classroom. Burke, however, at least in the early years, showed little bent for the practical. After he began his teaching career in 1943, he came to realize that teachers have special needs. He tells that he even began a practical book called *Devices,* but, in characteristic fashion, set it aside because he thought it needed a "preparatory grounding in the sort of work I did in my *Grammar of Motives* and *Rhetoric of Motives.*"[81]

In the absence of a text based entirely on Burke's theory, what uses of his ideas have been made? Perhaps the one that comes closest to holding together theory and practice is the use of the pentad as a heuristic device for purposes of invention because asking the questions—filling in the blanks—is an application of Burke's own epistemic scheme. The pentad is part and parcel of dramatism as a whole.

W. Ross Winterowd was perhaps the first to explain the pentad in his text *Rhetoric and Writing* (1965). In a chapter on "Invention: Building the Case," he introduces the terms and method for analyzing. He applies the method to a hypothetical newspaper editorial and at the end of the chapter asks students to "evaluate a source (magazine, article, editorial, telecast) by systematically applying the 'pentad' to it."[82] Winterowd, therefore, uses the pentad as a pedagogical device basically as Burke himself had used it in his criticism.

Winterowd's text goes beyond the application of the pentad. It is infused with Burke's ideas on definition and form. Winterowd even includes as readings Burke's "The Rhetoric of Hitler's Battle" from *The Philosophy of Literary Form* and "The Nature of Form" from *Counter-Statement*. Winterowd is making an effort to go "beyond the mere and often dull and sophomoric question of 'how to' into the more exciting questions of 'why?' " and "to achieve scholarly respectability within the framework of rhetoric as a discipline."[83] Unfortunately, the composition community of 1965 was not yet receptive to a book that openly proposed practice informed by theory. By 1975, when Winterowd restated the same objectives in *The Contemporary Writer* (Harcourt Brace Jovanovich), he met with a far greater sense of readiness in the academic community.

Rhetoric: Discovery and Change (1970) by Richard Young, Alton Becker, and Kenneth Pike remains the most complete rhetoric informed by a theory. And it is true to the spirit of a rhetoric text in the sense that it pays sparse attention to the conventions of manuscript form that are the usual marks of the handbook. A book like *Rhetoric: Discovery and Change*, which begins with a section entitled "Rhetorical strategies and images of man" and ends with a section entitled "Beyond analysis," which speaks of the need for a new rhetoric that has "as its goal not skillful verbal coercion but discussion and exchange of ideas," which quotes Burke's definition of rhetoric, expands his concept of form, and gives extensive treatment to the principle of identification, would seem to be a rhetoric primarily informed by Burke's thought. Actually, that was not the case. The assumptions of Kenneth Pike's tagmemic theory informed the book from its inception as early as 1964, and Burke's influence came later when the authors realized that many of Pike's propositions and Burke's ideas were consistent with one another, with the result that one could reinforce and enrich the other.[84]

Chief among the parallel ideas is Pike's theory of change and Burke's concept of identification. With a statement like the following from Pike:

> A shared unit implies the presence of a larger system embracing the two systems. The larger system constitutes an interaction matrix for the mixing.[85]

compare Burke's statement:

> Any specialized activity participates in a larger unit of action. "Identification" is a word for the autonomous activity's place in this wider context, a place with which the agent may be unconcerned.[86]

Or Pike's:

> A crucial concept in the model [of linguistic change] is that of shared component.[87]

with Burke's:

A doctrine of consubstantiality [cp. shared component], either explicit or implicit, may be necessary in any way of life.[88]

Both Pike and Burke begin with the function of language and ultimately find themselves considering cultural matrices that embrace all human thought and human relations. Pike began with a focus on heuristics as a device for the analysis of exotic languages; he later realized that the terms such as "particle, wave, and field" and "contrast, variation, and distribution" were "certain universal invariants" that "underlie all human experience as characteristics of rationality itself"—in short, a logic of thought.[89] Similarly, in the Introduction to *A Grammar of Motives,* Burke speaks of the terms of the pentad as "forms of thought which, in accordance with the nature of the world as all men necessarily experience it, are exemplified in the attributing of motives. . . . [The key terms] need never to be abandoned, since all statements that assign motives can be shown to arise out of them and to terminate in them."[90] Both of these approaches have universal implications.

The significant use of tagmemic heuristics in *Rhetoric: Discovery and Change* as a technique for analyzing human experience influenced William F. Irmscher in the first edition of *The Holt Guide to English* (1972) to extend the function of Burke's pentad and to formulate questions that would make it a generative model applicable to all situations, past, present, future, and probable.[91] The first edition of *The Holt Guide* thus laid the main claim to the use of the terms independent of Burke's original intent.

When Burke was invited to address a group of rhetoricians and composition specialists at a meeting of the Modern Language Association in 1977, he was asked to comment on the use of the pentad as a heuristic device. He immediately pointed up the difference between his application of the pentad and Irmscher's:

> My job was not to help a writer decide what he might say to produce a text. It was to help a critic perceive what was going on in a text that was already written. Irmscher uses "dramatistic" terms as suggestions for "generating a topic." My somewhat similar expression "generative principle" is applied quite differently.[92]

The key word in this statement is "text." Burke uses the pentad to analyze symbolic texts; Irmscher uses it, in addition, to analyze experience and, as preparation for writing, to bring to experience an ordered view, as one might imagine the dramatist imposes structure on life situations in order to create drama. Burke says, "Men have talked about things in many ways, but the pentad offers a synoptic way to talk about their talk-about."[93] Basically, Irmscher substitutes "experience" for "talk-about."

In explaining dramatism, Burke consistently makes a distinction between drama proper, the symbolizing or imitation of an action, and
"dramalike situations in real life." "Dramatistic" refers to a "critical or
essayistic analysis of language, and thence of human relations generally, by the use of terms from the contemplation of drama." Why does
he not include the contemplation of dramalike situations in real life?
Because they lack "finality" and "fixity." Burke insists he is examining
language, not reality—thus, the importance of written texts. Great dra-
mas, he says, are the "equivalents of the laboratory experimenter's 'test
cases.' "[94] They can be controlled and examined over and over again.
On this matter, Richard Chase has commented, "Nobody has ever
taken so literally the idea that all the world's a stage."[95] Knox characterizes imaginative works in Burke's terms as "stylized answers" to "questions posed in the artist's real life situation."[96]

Despite Burke's effort to differentiate between drama and dramalike
situations and to emphasize that he is concerned with drama as text, he
has insisted on more than one occasion that dramatism is not meta-
phorical, but literal,[97] to which one is prompted to respond, "Yes, it is;
and no, it is not"—a both/and-alternative that Burke allows as a way of
avoiding polarities. Dramatism is literal when Burke speaks of the symbol-using animal as actor, a phrase quite different from the figurative
implications of the symbol-using creature as automaton or the brain as
computer.

When dramatism represents a way of looking at reality, however, it
does become a metaphor. It imposes the order of literature on actual-
ity, which to Burke seems "so confused and complicated as to seem
almost formless, and too extended and unstable for orderly observation." Of course, confusion is very often the source of difficulty for
inexperienced writers. They need order in their thinking, and the pentad provides it. As a metaphor, it employs terms different from those
of the tagmemic metaphor, for instance. Each model may generate
comparable data, but the route—the metaphor—differs.

Burke leaves no doubt that he thinks written texts are capable of a
kind of conceptual clarity that other symbol-using arts, such as music,
painting, sculpture, dance, and architecture, are not. Yet, despite his
early belief that a literary text is the "perfect case" for dramatic analysis, Burke by 1955 was willing to extend the concept of text to the
products of various recording devices because they allow "repeated
analysis of a single unchanging development (an 'action' that, in its
totality, remains always the same)."[98]

The application of the pentad to real-life situations lacks the element
of replicability, except as memory supports the image of an action.
Examining each of the terms in detail, however, broadens the perspective of the observer and brings each of the terms into an interplay with

each of the others. To neglect any one is to fail to see the total drama of human relations. The purpose of a heuristic, as Richard Young explains, is to retrieve information, to examine situations from different perspectives, and by a systematic approach to encourage intuitive responses that make possible an hypothesis. Commenting on Irmscher's adaptation and extension of the use of the pentad as compared with his own intent and use, Burke generously conceded in his MLA speech: "Both uses have their place."[99]

One of the fundamental questions in assessing any theory is whether it works, not just for the author alone but for others as well. Must theories work to be sound? In a fairly recently translated essay by Kant entitled "On the Old Saw: That May Be Right in Theory, But It Won't Work in Practice," Kant indicates that theories must work. Yet if they do not at some particular time, that fact may not necessarily be invalidating. Time may be a factor in finding the application of a theory.[100] Certainly, time seems to have been a factor in the recognition of Burke's rhetoric by the composition community. Readiness was necessary before Burke could be read meaningfully in terms of composition. It was a scene-act ratio. In order for Burke's influence to become increasingly pervasive in the future, practitioners will need to reflect on questions such as these: Are efforts to apply Burke's dramatistic methodology bound by his original intentions? Can it be applied to new "texts," even non-texts? Can it be expanded to apply more specifically to new situations? Is borrowing of concepts from Burke without reference to the whole theory necessarily reductive? Does Burke's synoptic view fit compatibly with other views so that eclecticism can represent a coherent scheme and method?

Only in the last fifteen years have a number of key concepts from Burke infiltrated the practice and teaching of composition although one cannot claim that they have had a major impact. What then, in summary, are these?

1. The importance of terms to identify who and what a writer is.
2. The realization that terms provide screens, narrowing or broadening the context in which we think and operate.
3. An awareness of the generative possibilities of the pentad as a synoptic view of and way of talking about text and experience.
4. A preference of identification to persuasion as the aim of writing.
5. An understanding of the role of the negative in the principle of defining: What is it not?
6. The conception of form, not as a static mold, but as a viable process creating expectation and ultimate closure for the reader.
7. A realization that audience may be a matter of internal dialogue, the "I" addressing the "me."

8. The importance of dialectic—the linguistic give-and-take—as a way of finding solutions to seemingly irreconcilable opposites.

As valuable and useful as these insights may be to teachers of writing, however, Burke's major importance needs to be stated in different terms. He has provided composition specialists with a way of talking to one another conceptually about the act of writing in all its complexity. He has provided an epistemic base for the discipline. His synoptic views of rhetoric have added a heightened sense of purpose to the daily enterprise of teaching composition. By reaching the apex of the verbal pyramid he set out to construct for himself, he has established the claim of rhetoric as the all-summarizing discipline of human relations. Above all, Burke the man has become to many practitioners an ultimate figure, whose presence has brought new dignity and intellectual respectability to rhetoric and the teaching of writing. Now, in his octogenarian years, Burke continues to play his role as rhetor par excellence. In his typically disarming way, he tells us, "What I am proudest of is that I do the only thing I ever wanted to do. I just want to go on with the goddamn stuff as long as I can."[101]

Notes

1. *A Rhetoric of Motives* (New York: George Braziller, 1955), p. 22.
2. From *The Lion and the Honeycomb* in *Critical Responses to Kenneth Burke, 1924–1966*, ed. William H. Rueckert (Minneapolis: University of Minnesota Press, 1969), p. 245.
3. "Rhetoric—Old and New," *The Journal of General Education* 5 (April 1951), p. 203.
4. "Kenneth Burke and the Criticism of Symbolic Action," in *Critical Responses*, p. 212.
5. "Rhetoric—Old and New," p. 203.
6. (Boston: Allyn and Bacon, 1965), p. vii.
7. "Dancing with Tears in My Eyes," *Critical Inquiry*, 1 (September 1974), p. 27.
8. "Preface to the Second Edition," *Counter-Statement* (Los Altos, Calif.: Hermes Publications, 1953), p. xiv.
9. P. 265.
10. "Burke, Burke, the Lurk," in *Critical Responses*, p. 262.
11. "Prolegomena to Kenneth Burke," in *Critical Responses*, p. 247.
12. *Exile's Return: A Literary Odyssey of the 1920's* (New York: Viking Press, 1951), pp. 15–16.
13. Ibid., p. 24.
14. Preface to *The Complete White Oxen* (Berkeley: University of California Press, 1968), p. xi.
15. *Critical Responses*, p. 344.
16. "The Mark of a Poet: Marianne Moore," in *Critical Responses*, p. 400.
17. *Counter-Statement*, p. 215.

18. Ibid., pp. 214–215.

19. "The Technique of Mystification," in *Critical Responses*, pp. 89–90.

20. "The Sceptic's Progress," in *Critical Responses*, p. 53.

21. Preface to First Edition of *Towards a Better Life*, 2d ed. (Berkeley: University of California Press, 1966), pp. xxi–xxii.

22. "Prolegomena to Kenneth Burke," in *Critical Responses*, p. 250.

23. *Counter-Statement*, p. 210.

24. Ibid., p. 124.

25. "Prolegomena to Kenneth Burke," in *Critical Responses*, p. 249.

26. From *The Lion and the Honeycomb*, in *Critical Responses*, p. 246.

27. P. xiii.

28. *Collected Poems 1915–1967 by Kenneth Burke* (Berkeley: University of California Press, 1968).

29. "Foreword to *Book of Moments*," in *Collected Poems*, p. vii.

30. *Rhetoric*, p. 294.

31. "Prolegomena to Kenneth Burke," in *Critical Responses*, p. 248.

32. *Webster's Third International Dictionary* added a specific meaning of *dramatism* based on Burke's concept of the word: "a technique of analysis of language and thought as basically modes of action rather than as means of conveying information."

33. In *Language as Symbolic Action* (Berkeley: University of California Press, 1966), p. 3.

34. "Definition of Man," in *Symbolic Action*, p. 16. In response to current attention to the sexist implications of certain words, Burke now says "humankind" instead of "man."

35. *Rhetoric*, p. 20.

36. (New York: George Braziller, 1955), p. 319.

37. *Rhetoric*, p. 192.

38. "Linguistic Approach to Problems of Education," in *Modern Philosophies and Education*, ed. Nelson B. Henry, the Fifty-fourth Yearbook of the National Society for the Study of Education, Part I (Chicago: University of Chicago Press, 1955), p. 263.

39. "Definition of Man," p. 22.

40. Ibid., p. 21.

41. Letter to the author, 11 January 1984.

42. P. 23.

43. *Counter-Statement*, p. 42. See also "Linguistic Approach to Problems of Education," p. 267.

44. 2d rev. ed. (Los Altos, Calif.: Hermes Publications, 1954), p. 263.

45. *Grammar*, pp. 227–228. See also "Dramatism," *International Encyclopedia of the Social Sciences*, 7 (1968), p. 446. To the five basic terms, Burke later adds a sixth, attitude, which he defines as "incipient action" and includes as a subdivision of Agency. Agency is thus divided into means and attitude. See *International Encyclopedia*, p. 446; also "Addendum for the Present Edition," *A Grammar of Motives and A Rhetoric of Motives* (Cleveland: World, 1962), p. 443.

46. "A Dramatistic View of the Origins of Language and Postscripts on the Negative," in *Language as Symbolic Action*, p. 429.

47. *Grammar*, p. xi.
48. (Boston: Beacon Press, 1961), p. 26n.
49. *Rhetoric*, p. 38.
50. Ibid., p. 50.
51. *Grammar*, pp. 51–52.
52. *Rhetoric*, pp. 269, 271.
53. *Rhetoric of Religion*, p. 303.
54. *Rhetoric*, p. 277.
55. *Rhetoric of Religion*, p. vi.
56. Armin Paul Frank, *Kenneth Burke* (New York: Twayne Publishers, 1969), p. 19.
57. "Dancing with Tears in My Eyes," p. 26.
58. *Rhetoric*, pp. 293–294.
59. *The Philosophy of Literary Form*, 3d ed. (Berkeley: University of California Press, 1973), pp. 448–450.
60. *Rhetoric of Religion*, pp. 310–311.
61. "Linguistic Approach to Problems of Education," p. 302.
62. "A Dramatistic View of the Origins of Language," p. 431.
63. Ibid., p. 435.
64. Ibid., p. 421.
65. *Rhetoric of Religion*, p. 187.
66. "Linguistic Approach to Problems of Education," p. 243. See also "On Human Behavior Considered Dramatistically," Appendix to *Permanence and Change*, pp. 274–294.
67. (Berkeley: University of California Press, 1982), pp. 288–292.
68. On April 30, 1981, Burke received the National Medal for Literature at the American Book Award ceremonies. In March 1984, Temple University and the Speech Communication Association sponsored the "Burke" Conference, celebrating Burke's more than sixty years of scholarship and creative work. Burke appeared as critic-at-large.
69. *The Journal of General Education*, 5 (April 1951), pp. 202–209.
70. Letter to the author, 11 January 1984.
71. (New York: Russell & Russell, 1959), p. 128.
72. *Roots for a New Rhetoric*, p. vii.
73. Ibid., p. 57n.
74. Ibid., p. 120.
75. Ibid., p. 126.
76. P. 26.
77. *Critical Responses*, p. 182.
78. *Rhetoric of Religion*, p. 39.
79. In *Language as Symbolic Action*, p. 45.
80. *Rhetoric of Religion*, p. 299.
81. "Dancing with Tears in My Eyes," p. 26.
82. *Rhetoric and Writing* (Boston: Allyn and Bacon, 1965), p. 180.
83. Ibid., p. ix.
84. I am indebted to Richard Young, who in a letter dated October 20, 1983, describes the genesis and development of *Rhetoric: Discovery and Change*

(New York: Harcourt Brace Jovanovich, 1970) and points out parallels between Pike's ideas and Burke's.

85. "Toward a Theory of Change and Bilingualism," *Studies in Linguistics*, 15 (1960), p. 2.

86. *Rhetoric*, p. 27.

87. "Stimulating and Resisting Change," *Practical Anthropology*, 8 (1961), p. 267.

88. *Rhetoric*, p. 21.

89. "Beyond the Sentence," *College Composition and Communication*, 15 (October 1964), p. 129.

90. P. x.

91. (New York: Holt, Rinehart and Winston, 1972), Chapter 4.

92. "Questions and Answers about the Pentad," *College Composition and Communication*, 29 (December 1978), p. 332.

93. *Grammar*, p. 56.

94. "Linguistic Approach to Problems of Education," pp. 264–265.

95. "Rhetoric of Rhetoric," in *Critical Responses*, p. 253.

96. George Knox, *Critical Moments: Kenneth Burke's Categories and Critiques* (Seattle: University of Washington Press, 1957), p. 35.

97. For various references, see "Dancing with Tears in My Eyes," p. 28; "Linguistic Approach to Problems of Education, pp. 259–260; *International Encyclopedia*, p. 448.

98. "Linguistic Approach to Problems of Education," p. 265.

99. "Questions and Answers about the Pentad," p. 330

100. Trans. E. B. Ashton (Philadelphia: University of Pennsylvania Press, 1974), pp. 41–43.

101. Richard Kostelanetz, "A Mind That Cannot Stop Exploding," *The New York Times Book Review*, 15 March 1981, p. 26.

Selected Publications

For the new reader of Burke, certain works are better starting points than others simply because Burke's style is more accessible in these pieces. Chief among these is the article on dramatism that Burke himself wrote for the *International Encyclopedia of the Social Sciences*, Volume 7 (1968), pp. 445–451.

Because Burke writes for a new audience of educational theorists and senses the need to explain the basis of his own theories, his article "Linguistic Approach to Problems of Education," in *Modern Philosophies and Education* (Chicago: University of Chicago Press, 1955), pp. 259–303, is particularly helpful for the beginner. The essay also includes a long bibliographical note written by Burke, commenting on his own works and selected pieces written about him.

In *Language as Symbolic Action*, Burke groups five essays under the title of "Five Summarizing Essays," because they "best convey the gist of the collection as a whole." Especially important among these before reading the major works is "Definition of Man," pp. 3–22.

All eleven of Burke's major works and collections have now been reissued by the University of California Press, Berkeley, many of them with new prefaces by Burke. The list includes:

The Complete White Oxen: Collected Short Fiction
Counter-Statement
Towards a Better Life, Being a Series of Epistles or Declamations
Permanence and Change: An Anatomy of Purpose
Attitudes Toward History
Language as Symbolic Action: Essays on Life, Literature, and Method
The Philosophy of Literary Form
A Grammar of Motives
A Rhetoric of Motives
The Rhetoric of Religion: Studies in Logology
Collected Poems, 1915–1967

William H. Rueckert's *Kenneth Burke and the Drama of Human Relations,* 2d ed. (Berkeley: University of California Press, 1982), includes a full bibliography, pp. 295–317, listing books and articles by Burke and selected reviews, essays, and works about Burke. The section on secondary sources is annotated.

Absolutely invaluable to the serious student of Burke is William H. Rueckert's edited collection entitled *Critical Responses to Kenneth Burke, 1924–1966* (Minneapolis: University of Minnesota Press, 1969). It also includes both an extensive bibliography of the writings of Burke, pp. 495–512, by Armin Paul Frank and Mechthild Frank and an additional bibliography of works about Burke, pp. 515–522, by William H. Rueckert.

The following selection of articles pertaining specifically to rhetoric and composition supplements Rueckert's entries:

Burke, Kenneth. "Questions and Answers about the Pentad." *College Composition and Communication,* 29 (December 1978), pp. 330–335.
    Burke's response to the use of the pentad as a generative device and his re-explanation of his original intentions.
Comprone, Joseph J. "Burke's Dramatism as a Means of Using Literature to Teach Composition." *Rhetoric Society Quarterly,* 9 (Summer 1979), 142–155.
    Explains in detail a rhetorical-dialectical heuristic that can be applied to literature as a way of teaching students to write critical essays.
———. "Kenneth Burke and the Teaching of Writing." *College Composition and Communication,* 29 (December 1978), pp. 336–340.
    Applies the basic terms of the pentad to stages of composing, establishing that writing is an actualization or dramatization of thought.
Feehan, Michael. "Kenneth Burke's Discovery of Dramatism." *Quarterly Journal of Speech,* 65 (1979), pp. 405–411.
    Traces the grounding of Burke's dramatism in a short article entitled "Twelve Propositions by Kenneth Burke on the Relation Between Economics and Psychology," reprinted in *Philosophy of Literary Form* in an abbreviated version.
Heath, Robert L. "Kenneth Burke on Form." *Quarterly Journal of Speech,* 65 (1979), pp. 392–404.
    Traces the evolution of Burke's concept of form.

Irmscher, William F. "Analogy as an Approach to Rhetorical Theory." *College Composition and Communication*, 27 (December 1976), pp. 350–354.

Uses an extended version of the pentad to show correspondences between the rhetoric of writing and the rhetoric of other performing arts.

Keith, Philip M. "Burke for the Composition Class." *College Composition and Communication*, 28 (December 1977), pp. 348–351.

Uses particular Burke techniques to encourage students to see familiar experiences in new ways.

———. "Burkeian Invention, from Pentad to Dialectic." *Rhetoric Society Quarterly*, 9 (Summer 1979), pp. 137–141.

Suggests how a number of strategies Burke uses in *A Grammar of Motives* can be taught to freshman students as a way of giving them more sophisticated approaches to invention.

Kneupper, Charles W. "Dramatistic Invention: The Pentad as a Heuristic Procedure." *Rhetoric Society Quarterly*, 9 (Summer 1979), pp. 130–136.

Places special emphasis on the ratios as the generative potential of the pentad.

McCrimmon, James M. "Will the New Rhetorics Produce New Emphases in the Composition Class?" *College Composition and Communication*, 20 (May 1969), pp. 124–130.

Burke and Richards are the points of departure for a discussion of what can be gained from the new rhetorics for teaching composition.

O'Keefe, Daniel J. "Burke's Dramatism and Action Theory." *Rhetoric Society Quarterly*, 8 (Winter 1978), pp. 8–15.

Examines the limitations of Burke's distinction between action and motion.

Schwartz, Joseph. "Kenneth Burke, Aristotle, and the Future of Rhetoric." *College Composition and Communication*, 17 (December 1966), pp. 210–216.

An analysis of likenesses and differences between Aristotle and Burke on the premise that what is the "new rhetoric" will have validity only if it is an integral part of the vital tradition of the "old rhetoric."

Smith, Nelson J., III. "Logic for the New Rhetoric." *College Composition and Communication*, 20 (December 1969), pp. 305–313.

Burke is the point of departure for a discussion of a logic for classroom use.

Turner, Linda M. "On First Reading Burke's *A Rhetoric of Motives*." *College Composition and Communication*, 24 (February 1973), pp. 22–30.

A young woman of the 1970s identifies with the concepts of Burke's *Rhetoric* and finds a synthesis to her own multiple interdisciplinary interests.

Winterowd, W. Ross. "Dramatism in Themes and Poems." *College English*, 45 (October 1983), pp. 581–588.

Distinguishes between propositional (linear) and appositional (nonlinear) writing. Advocates teaching the "representative anecdote" from *A Grammar of Motives* as a means of development.

THEODORE BAIRD

by Walker Gibson

> No account that I have ever made conveys what the course
> is like day by day and in the students' papers. It is proba-
> bly both better and worse than you think.

Theodore Baird was born in 1901, graduated from Hobart College in
1921, and in 1929 received his doctorate from Harvard. After brief
tenures at Western Reserve and Union College, he began his long
career at Amherst in 1927. For the next forty-three years he engaged
in the varied chores required in a small department with a strong
literary tradition, including a very popular course in Shakespeare; during most of these years he occupied the Samuel Williston Professorship.
His reputation today, however, rests not so much on his teaching of
literature as on his leadership in the teaching of freshman composition.
For a quarter century Baird was director of English 1–2, the yearlong
required course in freshman writing. It is for this course that he himself would want to be remembered, and it is this course and its development under his guidance that concern me in what follows.

In the mid-1960s, English 1–2 was finally replaced by other approaches to the freshman requirement, as inevitably happens to all
such enterprises. The feelings Baird expressed on that occasion, to one
of his many correspondents, say something about the quality of his
commitment over the years of his directorship:

> The other night English 1–2 was voted out of existence by the Faculty,
> and of course I have been thinking about this and the past and trying to
> see where I move in the remaining years. This business has been going on
> for 25 years, and no one, except a few people, you included, can guess
> how much it has occupied my attention, how much fun I have had, what a
> privilege it really has been, to be allowed to try to set up teaching in these
> many ways. I wonder if you don't remember the best period when despite
> much friction supplied by my own irritable nature much was being clari-

fied for us as a group, when week by week discoveries were being made. I don't think I ever had a clearer motive than the determination not to be bored and everything as far as I was concerned followed from that. Of course this was asking more than one has a right to do, but I used to be true to my feelings—with very few compromises. I suppose I ought to recognize this kind of activity is all over for me . . . Yet I swear the assignments I made last year were the best and most amusing I ever made. Can I be wrong about this? Do not answer.

And again, some time later:

The real issue, as I see it, is this: students say that the possibilities of Life are endless, and I say, education should describe at least some of the limitations to Life. Such as: you do have a body, you exist in time, you are often generally wrong. They say, You are old Father Theodore, and the hell with you. . . . I suppose what I wanted was simply to demonstrate that the teaching of this course was as engrossing as the teaching of an advanced course, that you could make a Life out of it as well as by being a Wordsworth man. This is where I failed with these people [younger colleagues]. They have been discussing the future of Freshman English (when I go) and no one comes up with the promise that he will really work at it. Their thinking can be no better than anyone else's when they devote no more than others do elsewhere.

What kind of course was it that could arouse all this energy and this passion? The fact is that no descriptive account of English 1–2 has ever been published other than a few documents for local consumption by Baird himself. Indeed, though the course enjoys a certain vague reputation and is known to have influenced a number of college teachers, it has received precious little professional notice. No book ever emerged as an official text of the course; no grant was ever awarded; no session of NCTE or CCCC, so far as I'm aware, was ever concerned with English 1–2 at Amherst College. When one recalls all the solutions to the freshman English problem that have emerged in recent years, from sentence-combining to tagmemics, together with the considerable hype that has accompanied each, the Amherst answer seems becomingly modest, if not oddly diffident. There is another side to this. There was some smugness in the Amherst position that advertising to the wide world would do no good. How could one make people out there really understand? Furthermore, the course is an ongoing activitythe argument went, not a body of knowledge to be readily laid out on paper. It is an action.

Granting the argument, I make here an effort to sketch in admittedly static print something of the character of that action. Two things to clarify at the outset have to do with format or structure. First, the course was team-taught in the sense that every instructor was expected to take part in its planning and development at every stage. Second, the course offered its students an integrated sequence of assignments (different each year) that began somewhere and more or less logically led somewhere else.

The course was team-taught. This does not mean that a group of people met a few times and went their several ways, nor that they participated directly in one another's classrooms. In English 1–2, teamwork meant something else and something fairly rigorous. A draft of assignments was composed over each summer by a member of the staff (often Baird himself); this draft was then discussed, amended, agreed to in a trial form. Staff members (eight or ten individuals, that is, a majority of the department) met weekly during the academic year for a couple of hours or so, and absence from these meetings was unthinkable. The agenda for every meeting was the same: how to edit and rephrase the next week's assignments so as to elicit some worthwhile response from students.

Baird's faith in this form of teaching-by-consensus seems to have originated in the 1930s, when with four other young teachers he participated in an earlier composition course known as "English 1C." An article he wrote for the local alumni magazine in 1939 concludes:

> The only virtue of English 1C is that five teachers are engaged in a common effort to see to it that the freshmen do as much writing as we can read and that the writing is as decent, as clear, as sensible, as intelligent as we know how to make it. And the five teachers come together to make one course, not for the appearance of uniformity but because by an exchange of ideas, by self-criticism, by argument, we can define our objects more clearly and use the best methods for achieving them that we know about.

It should be added that the cooperative give-and-take Baird describes as English 1C had to have been different in important ways from the later cooperative efforts that became English 1–2. It is one thing for five young peers to meet and agree on a common procedure. It is quite another thing to meet with twice as many who range in sophistication and experience from the newest departmental recruit to the senior director himself. And in English 1–2 there was a senior director, and he did direct, though he also made conscientious efforts to involve everyone democratically in the decision making. There are, of course, limits to the uses of democracy in educational planning. Dull people meeting in equality together will produce dull materials unless led by an individual with energy and imagination. In English 1–2, no one doubted where the energy and imagination were finally coming from, and no one doubted who was in charge, for all the genuine cooperation and consensus that did go on.

In any event, for the dozens of young instructors who passed through this regimen (and I was one), the experience was heady in the extreme. Most of these instructors (though not I) labored under a three-years-and-out-you-go rule, which was cruel treatment, but most of them left feeling that their minds had been permanently changed—and for the better. More of that later.

The second feature of the course into which Baird put all the strength of his convictions concerns the assignments themselves. First, there was the sheer number of them, over thirty per semester. Directions for writing a short paper (one to three pages) were distributed at every class meeting, due at the next class. These papers were read, commented on (without grades), returned to the students and discussed at the following meeting of the class. (The teaching labor may sound intolerable, but it wasn't, not quite. Beginning instructors taught three sections of from twenty to twenty-five students each. Others taught a single section as part of a normal three-course load. Only one instructor, to my knowledge, actually cracked up.) The schedule of assignments was followed precisely by all sections of the course, which meant that over any given 48-hour period the entire freshman class (300 young men with outrageously high SAT scores) was engaged in solving the same set problem. That this resulted in a good many dormitory seminars among students was inevitable and probably beneficial.

More important, the agreed-on assignments were sequential: they led from one place to another. During the first semester (English 1), they were concerned with some aspect of the students' nonacademic or extracurricular experience; in English 2, attention was directed to their classroom experience, specifically in history and in science. No text or reader was ever used; selections from the students' written work, mimeographed and distributed, became what text there was. The electronic copier, of course, was not then available; it would have been invaluable.

Sets of assignments, particularly in English 1, were organized around some concept or *key term.* One early year the term was *technique,* and the main question was, How can you say how it is that you do something you know how to do? (The conventional label "process paper" was never mentioned.) How can you describe in words the special feel of a successful serve in tennis? Answer: You can't. What can you describe? An order of actions? Baird used this example in a piece he wrote about the course in 1952, addressed to parents and alumni:

> A student who is a good tennis player sets out to write a paper on what he does when he serves a tennis ball. He knows he knows what he is writing about, yet as he begins to address himself to his subject he immediately encounters the inescapable fact that his consciousness of his own action contains a large area of experience quite beyond his powers of expression. The muscular tensions, the rhythm of his body as he shifts his weight, above all the feel of the action by which he knows a stroke is good or bad,

all these and much more lie beyond his command of language, and rendered almost speechless he produces a mess. He knows in the sense that he can perform the action, but he does not know in the sense that he can communicate this action to a reader. At this point the teacher tries to get him to distinguish between these two levels of experience, to become aware of them, to generalize about them. The next step is for the student to take for himself by recognizing that a part or an element of his experience can be communicated to another person when he isolates the order in which he throws the ball into the air, raises his racket, and so on, and that the order of his actions as distinguished from the action itself is the subject of his writing. The student may even perceive that between the order of movements as he sees them and the order of words in a sentence some relation can be made, and that when he has made this relation he knows what he is talking about.

Other organizing themes in other years included the experiences of puzzle-solving, of playing games (What is a game?), of reading a road map. Glancing at such a list, it is perhaps no wonder Amherst people despaired of making the course appear serious and sensible to the outside world, and what the outside world picked up about English 1–2 did Amherst's reputation no good. "Oh you teach at Amherst," a professor at a neighboring college once said to me. (My years there date from 1946 to 1957.) "You teach at Amherst, do you? That's the place where you tell your freshmen how to read a road map." This was said contemptuously. And yet I remember fondly that sequence on the road map. Students were issued an oil company's map—and other maps too—a textbook for once, and assignments were generated around that text. "What is a road map?" we finally asked, and the answer (of course) was that it is a set of directions, a guide for certain kinds of human action. It is not an expression of terrestrial reality. Many weeks of effort were required to arrive at this revelation, commonplace though it sounds. Meanwhile our neighboring colleges thought us frivolous and silly, and no doubt we sometimes were.

The map is not the territory, the great semanticist Alfred Korzybski has memorably told us. We did not mention *Science and Sanity* to our students, but Korzybski's dictum underlay much that went on in English 1–2. The word is not the thing.

II

To illustrate a sequence of assignments in more detail, I turn to a set used in English 1 in 1963, long after I had passed from the scene. This is the very set Baird alludes to in his letter quoted earlier: "the best and most amusing I ever made." Whether that is true or not, this set bears a special interest because its subject, for a change, is one every composition teacher confronts every day. The subject is English usage, a uniquely academic-sounding concern for Baird and this course. How did Baird go about addressing the problem of English usage? He did it, characteristically, by *putting it to the students.* He did it by *complicating their lives,* not simplifying them. And he did it with a kind of mock-academic jocularity that for some people, anyway, brought a breath of fresh air into the stuffiest classroom.

In the archives of the Amherst College Library, the assignments for English 1, 1963 version, are filed in seventeen closely typed pages, in a large cardboard box crammed with other sets for other years. The length is typical and suggests that I will do less than justice to the sequence in a few paragraphs here. Nevertheless, though skipping and skimming, I can give the reader, I think, some sense of what a freshman went through—or was supposed to go through—in experiencing these assignments.

Baird began this sequence by quoting a university's advertisement he had seen in *The New York Times,* an ad offering an extension course called "Master Good Writing."

> . . . a thorough study of the mechanics of English—grammar, punctuation, sentence structure, diction, usage. . . . By the end of the course the student should be confident of his ability to meet formal standards of "correctness."

What Baird immediately noticed in that ad were the quotation marks. The student was asked:

> Why do you suppose "correctness" is printed with quotation marks around it? Do you see some significant difference between correctness and "correctness," in the context of that sentence? If you were asked to choose between an ability to meet the formal standards of correctness or "correctness," which would you choose and why? Or do you prefer not to choose? Why?

I assume that, in trying to answer these questions, the students floundered and fell all over themselves. Such floundering in the early stages was expected and indeed necessary. (If the students had been able to answer such questions wisely during the first week of the course, there would be no need to continue.) Yet we can see that Baird is here immediately pressing a fundamental issue about writing that he will relentlessly pursue in following assignments. Why "correctness" in quotes? Because even that ad writer knew, perhaps with embarrassment, the limitations to the kind of mastery one would learn from a course like "Master Good English." This is the mechanics of English, and that's something, of course, but that's all it is. As for *correctness,* without quotation marks—correctness is what good writers actually write! And they do so—this is surely the point—by constantly playing off the "formal standards" against their own voices and impulses, their sense of audience and purpose and who knows what. To be "correct" is easy; to be correct is to face all the vicissitudes of expressive life. It is something like that conflict, I take it, that Baird is inviting his students to recognize and illustrate. And, I further assume, they didn't get it.

An assignment immediately following asks them to develop this conflict of correct vs. "correct" in relation to clothing, manners, vocabulary.

> You have doubtless seen in these few days at college more than one person torn by this conflict, acting, as we say, unnaturally. Choose such an incident that you think you can write about. . . . Do it as a dramatic episode, that is, as an event with persons performing actions and speaking words in a particular setting.

The student must then perform variations on this exercise from other perspectives, with a first-person-singular narration and with an editorial generalizing on conflicts between the desire to conform and the desire to be independent. Of the editorial, the student is asked, "Why was this writing easy to do? What was 'correct' or correct about what you have just written?" Presumably, in writing an editorial, one finds it easy to slide into conventional "correct" ways of dodging the true, the agonizing conflict.

The issue of what good writers actually do with "correctness" is faced dramatically with an unbuttoned passage from *Catcher in the Rye*. The instructions read:

> It can be said that there is an incorrectness or "incorrectness" that makes it correct or "correct." Point out signs of this art. What can be meant by saying it makes incorrectness or "incorrectness" correct or "correct"?

It would be fun to know whether by this time some students are beginning to get the hang of all this. But if they do, Baird immediately turns them around, gives them pause.

> Suppose the style of J. D. Salinger were presented to you as a model to emulate. Why should such writing be held up to you as a model? Is this an English of limited use? Could Holden Caulfield get into medical school? Into Amherst College?

At this point (Assignment No. 10), the student is given a chance to pause and review, to "see how you can make a coherent statement about the problem we are considering."

> What is the problem exactly? What words have we been using to talk about it? Where do you see yourself in relation to the problem? Are you very clear in your own mind? Who or who else do you think is clear about this problem?

The student was praised, I conjecture, who admitted that he was anything but clear in his own mind.

Baird now confronts his students with some authorities on writing and usage. ("Who else do you think is clear about this problem?") The first is Emerson, who takes very high ground. "A man's power to connect his thought with its proper symbol, and so to utter it, depends on the simplicity of his character, that is, upon his love of truth and his desire to communicate it without loss." There follow in succeeding assignments statements on usage from Adams Sherman Hill (1892), from John F. Genung (1891), and from a 1960 college handbook. All these take fairly low ground. "The decisions of good use are final" (Hill). "The writer must see to it he keeps his mother tongue unsullied . . . transgressions of the standard are owing to want of culture" (Genung). The handbook counsels the student to ask, "What is being done at the present time by people of education, of taste, of social importance, people whose opinion I value?" Money is only just not mentioned. In each case, Baird asks what kind of world is created by the assumptions in these passages, and one of his points is that such worlds are very much still with us and around us.

> Where do you find this world when you look around you in Amherst?
> Does this discovery do anything for you?

(Amherst College in 1963, it should be remembered, was still dominated by prep-school WASPS, by fraternities, by faculty members educated at Amherst and Harvard who had their ideas of refinement and gentility. Genung himself was an Amherst professor, and his notion of how Amherst men should speak and write—PURITY he called it—was clearly felt to be the appropriate language for a privileged community. There are those who still think so. We might further remember, now that I'm in a parenthesis, that there were no women at Amherst in 1963; they were not admitted for another dozen years. In hindsight this was a handicap for English 1–2, for women bring their own insights to problems of propriety, problems of "correctness" and correctness. I venture to say that women students might have responded better than men to some of Baird's challenges.)

But if the student is about to conclude that elitist dialects like Genung's PURITY are all around him and should be resisted in the name of free spirits, Baird will not let him rest there. The next passage is from Robert A. Hall, exposing in all modern linguistic fervor our snobbishness about standards. "The real reason behind condemning somebody [for making a 'mistake'] is the desire to put that person in his or her place." But, asks Baird:

> What is your English teacher, or any other teacher, doing when he "corrects" your grammar? Or your diction? Or your spelling?

My respected teachers are simply putting me in my place, are simply acting out snobbish prejudices? Or is Genung right after all? The students' heads must be whirling.

We next take up that last-mentioned pedagogical obsession, spelling, with a long and informative quotation from Harold Whitehalls's *Structural Essentials of English*. Now Baird makes another distinction. There are misspellings, he says, that give "a certain pleasure," and he quotes a long and hilarious list allegedly lifted from freshman papers: clumbsy, Rose Bowel Game, to all intense purposes. On the other hand, the common misspellings (recieve, occured) are only annoying. "Put into words the conflict of feelings you have about conflicting standards about spelling and misspelling."

All this is quite consistent with what has gone before. There are misspellings that are wonderful, like Joyce's, even when they're inadvertent. (Every teacher keeps lists of these.) But the dumb mistakes (its for it's) are just dumb mistakes. The student is, therefore, encouraged to see once again that there are times when "incorrect" can be correct, other times when "incorrect" is indeed incorrect.

With Assignment No. 21, we have another chance at reconsideration and assessment:

> Reconsider the last five assignments dealing with the standards of correctness and the meaning of these standards in terms of people and society. The process of moving from one assignment to the next may be confusing, but this usually happens when you try to place yourself historically by describing the thought and feeling and behavior of other people in the past or in the present. You are not the first person to find conflicting standards existing, as it were, side by side. If you find survivals of the past around you this is to be expected. You are not the first to live with confusion. Express as well as you can your sense of conflicting standards in their largest sense. Do this not only in terms of English usage but also in terms of the life you see around you . . .
> Finally, express what it means to "live with confusion." This is an assignment that makes demands on you.

The next few assignments ring the changes on distinctions between saying, "I was confused" and "I live with confusion." We turn then to situations when one is not confused at all, when one is *clear*. Examples abound, and one example where the invention of systems has made life clear is the experience of going to the library and finding a book. (Instructors offered here a little useful education on the library and the Dewey Decimal System, always a "unit" in English 1.) All is not lost after all; all is not chaos; we do have ways of ordering our lives and they can work.

> When you say of your room or of your desk, This room or desk is a mess, what do you mean? . . . You can say that mess equals chaos. Describe

chaos, using your room or desk as an example. But by writing this paper, you have made an order out of chaos. Explain.how you were able to do this.

The final assignment in the series (No. 32) reads as follows:

> Look back to the early assignments in this course on Correctness and Standards of usage and then reconsider the latest assignment on the operation of finding a book in the Library according to the D.D.S. and how this operation can be seen as complex. It is unlikely that anyone will now experience a sudden revelation and settle once and for all the question what Good English really is.
>
> Nevertheless it can be said that the students and teachers have been concerned with what may be called an Ostensible Subject for this course, when all the while we really are thinking about its Real Subject. How do you phrase this Real Subject? You must understand that this Real Subject exists only as you think of it. The question is how to express what everyone may know and yet knows differently.

Plenty of real subjects (Conflicts? Messes and orders? The human nature of communication?) should have resulted from this assignment. Notice how, as usual, the burden is placed on the student to come up with his own summation—"as you think of it."

There follows a so-called long paper (perhaps from six to ten pages), which I omit in the interests of pressing on to the final examination, which is a fairly straightforward reprise of the semester's themes. The exam quotes a commencement address at Amherst delivered in 1905 by a certain President Harris:

> The purpose [of college education] is to turn out refined, honourable men. The educated man is the all-around man, the symmetrical man. The one-sided man is not liberally educated. The aim of the college is not to make scholars. The aim is to make broad, cultivated men, physically sound, intellectually awake, socially refined and gentlemanly, with appreciation of art, music, literature, and with sane, simple religion, all in proportion. Not athletes simply, nor scholars simply, nor dilettantes, not society men, not pietists, but all-around men.

The students were then asked the following questions:

> Obviously this Ideal Educated Man would speak and write a good all-around English without solecism or errors in taste.
>
> What are the assumptions behind President Harris' remarks about the nature of people and of their society?
>
> Do you find these assumptions or some of them still alive and effective in the Amherst College of 1963? Where? How are they manifest?
>
> If you find President Harris' language inadequate or even offensive, what words and phrases and sentences would you yourself use in describing your own ideal of Amherst College?

The trouble with Harris, I take it, is that he was not living with confusion, and the student was certainly invited, in that last sentence of the instructions, to see Harris' language as "inadequate or even offensive." (Not difficult, of course, for any commencement address, let alone one delivered in 1905.) I judge the student might ideally conclude his exam by saying something of his own life with confusion. I live with confusion, he might say, but I am not confused. I can use language. I can make order out of chaos. If called upon, I can write "correct" English, and sometimes, ordering the confusions I live with, I can even write correct English. Then I know what I'm talking about.

III

Perhaps the first thing that will occur to even the sympathetic reader of these assignments is that they are not for everybody. The unsympathetic reader will add that they're not for anybody. Certainly, they ask a good deal of the student, and Baird always took advantage of his advantage: namely, a captive audience of highly motivated, decently prepared, and reasonably smart young men. That he saw his deep responsibility in this situation goes without saying. The assumption he made, in the very tone of his assignments, was a generous one: You and I are intelligent adults, eager for education, and I take it for granted that you will be as concerned and excited and good-humored as I about this business we're embarked on. (E.g. English usage.) But was this a fair assumption? Not always, of course. In moments of gloom—and Baird allowed himself occasional attitudes of exaggerated gloom—he confessed his doubts about that assumption:

> I see plainly where I have gone wrong—if wrong it is—in assuming that our students are better than they are, that they can be talked to from the level of adult interest (whatever it may be), that the best ideas I have are not too good for them. Plainly a strong case can be made against me, for with this really nice set of assignments I had many students asking, What's the idea of this course? and without shame!

(That question, needless to say, has been asked by people other than students—and without shame. What *is* the idea of this course? How is it different from other composition courses in purpose or goal? One answer is that it isn't different at all. For what do we want of our students? A critical attitude toward language? A recognition that the world they live in is the world they express in words? That control of that world and of themselves depends considerably on their control of their own words? Things like that. In such formulations we can feel the insidious creep of "correctness," the blur of our pious abstractions. Still, this is the sort of thing we all believe. This is surely what all composition courses are all about, and English 1–2 was no different. "It is probably both better and worse than you think.")

But Baird never remains gloomy for long, and a few lines later he turns it around:

> You mustn't think I say *I* am wrong. I couldn't do it on any other terms.

At any rate, the fact seems to be that, for very many students, Baird's influence was crucial to their understanding of their education. For most of the instructors who passed through his hands, as I have said, the same was true. (A partial list of men who have worked with Baird includes a number still at Amherst—G. A. Craig, Benjamin DeMott, John Cameron, William Heath, William Pritchard—as well as a larger number who have taken some part of Baird with them to other pastures—Reuben Brower, C. L. Barber, W. V. Clausen, Julian Moynahan, Roger Sale, William R. Taylor, Jonathan Bishop, John F. Butler, William Coles, myself.) Yet it is a curious sort of influence; it cannot readily be documented; it exists largely by word of mouth. Contrast the prestigious bibliography of a Richards or a Burke. Baird's publications, as I shall point out shortly, are few and brief, and they reached small audiences. How is it, the reader may reasonably be asking, that a man with a reputation seemingly so evanescent has won his way into this collection of distinguished names in rhetoric and composition?

The answer is worth telling as symptomatic of the way Baird's fame has flourished. Our editor, John Brereton, took his graduate work at Rutgers, whose English Department is well stocked with Amherst-trained professors. Richard Poirier, T. R. Edwards, and James Guetti are all Amherst graduates; Julian Moynahan taught at Amherst for some years. And in that setting Brereton heard the word, though, as he says, it was certainly at secondhand. It was perfectly logical for him, however, as he assembled his significant figures, to include that of Theodore Baird, and I here record my acknowledgment of his good sense.

An anecdote about Edwards is worth adding here. The policy during Baird's years with English 1–2 was to excuse almost nobody from the freshman requirement. If you came to college with an SAT verbal score of 800, you took the course anyway. But Tom Edwards entered Amherst with credentials so spectacular that even Baird relented, and Edwards proceeded forthwith into sophomore work and four years later graduated summa cum laude. But somewhere along the line—perhaps about junior year—he realized what he had missed, and he spent a semester sitting in the back row of Baird's English 1 section, making up for the omission.

I do not mean to suggest that Rutgers or any other institution has adopted Baird's methods or vocabulary in any systematic way. But many of us have adapted his approaches in various guises to very different audiences, suggesting that his effectiveness and relevance are by no means restricted to a "select" group of students. The most conspicuous example is William Coles at the University of Pittsburgh, whose teacher-training programs and several publications reflect in his own individual style attitudes he first learned at Amherst. (In fact, readers of Coles's *Composing* [Hayden, 1974] will find a version of the usage sequence described here, including the Emerson-Hill-Genung triple play and some other familiar passages.) I will also mention my own *Seeing and Writing* (Longmans, 1959 and 1974), a sequence of assignments that owes much to my Amherst experience.

But not everyone was favorably impressed by English 1–2, and there were those, both students and faculty, who were thoroughly hostile. Many, of course, didn't "understand." Baird's good friend Robert Frost was one of these—he said it was "kid stuff"—but then, as Baird ruefully remarked, we couldn't undertake to educate Frost. Some faculty members saw the course, quite rightly, as a threat to their own way of doing things. (Here are the facts, boys; learn them.) An important exception was the physicist Arnold Arons, director of Science 1–2. It is not too much to say that freshman English and freshman science enjoyed for some years a rapport and a common intellectual approach that must have been unique in American education.

And then the assignments themselves seem to cry out for parody, friendly or unfriendly. They were often burlesqued in one student publication or another. The course figures largely in Alison Lurie's *Love and Friendship* (1962), an irreverent novel about faculty life at "Convers College."

> Assignment 11
>
> Here is a photograph, an airview of Convers College,
>
> (a) Let us assume you are now somewhere in the middle of the area contained in the photograph and you recognize this as a photograph of the spot you are now on. What do you do to recognize this?
>
> (b) Define, in the context of (a) "the spot you are now on."
>
> (c) What difference do you see between this spot and the one in the map in Assignment No. 10?

If this assignment, evidently lifted from an actual series, looks far-out or silly in Lurie's novel, it was anything but silly in the context of the series, where the student was invited to see that "the spot you are now on" can be both anywhere and nowhere.

Actually, Baird's assignments are hard to parody successfully because

their language is so modified by self-ironies and admissions of final
failure that take the ground out from under the parodist. If you ask
questions to which you do not know the answers, you are relatively
invulnerable to a burlesque that does the same thing.

IV

"This freshman course has been the center of my entire intellectual life,
and more than that, for to it I have brought whatever I have learned as a
human being." So wrote Baird to the college president toward the end of
his teaching career. The question arises, What were the origins of English 1–2? What had Baird encountered in his reading or experience—
"as a human being"—that could account for the particular character of
the course? I certainly can't presume to answer that with any assurance.
Baird himself has recently remarked, "I would say that all my teaching
was in rebellion against my own formal education and my elders." I *have*
mentioned Korzybski, certainly an important influence, but there must
have been hundreds of others in and out of books, for Baird has always
been an omnivorous reader as well as a sharp observer. Occasionally, he
distributed a list of suggested reading to his staff, and I don't suppose
anyone actually looked up all those things—I know I didn't—but the lists
did at least suggest what *he* was reading. They are not the sort of lists we
hand out nowadays to our graduate seminars in rhetoric and teaching.
One, dated 1946, for instance, in addition to Korzybski, includes three
titles by the physicist Bridgman and two by the historian Collingwood.
William James is prominent. There are a number of books on how the
mind thinks: Polya, Wallas, Wertheimer. And there are also titles to
remind us that Baird was acutely aware of what was just getting under
way in linguistics and the study of grammar: Fries, Hayakawa, Marckwardt, Ogden and Richards.

A later list (1954) begins with E. D. Adrian's *The Physical Basis of
Perception* and ends with J. Z. Young's *Doubt and Certainty in Science*.
Take that, you young English teachers. In between appear again Bridgman, Collingwood, James, Korzybski, several linguists. Among current
titles there is one that made a big hit at Amherst at that time, McLu-
han's *The Mechanical Bride*.

To such lists ought to be added one name possibly omitted as too
obvious: Henry Adams. At one period (before my time) the entire
freshman course was organized around a close reading of *The Education*. That book was surely formative in Baird's own education, and one
can perhaps feel in his own prose style some echoes of Adams.

Authors such as those I have mentioned, together with the general
procedure of the course, have suggested to some that there must be a
specific intellectual source to all this, perhaps even a *secret* somewhere.

In one of the very few published critiques of English 1–2 that exist, James H. Broderick undertook to find such a source in early twentieth-century American philosophy. ("A Study of the Freshman Composition Course at Amherst: Action, Order, and Language," *Harvard Educational Review*, 28:1 [Winter 1958].) Broderick mentions, quite plausibly, logical positivism, operationalism, pragmatism, John Dewey's instrumentalism, and, of course, semantics. No doubt there is a lot to it; no doubt connections can be drawn. But it's important to recognize (as Broderick did recognize) that such language was not part of the vocabulary of the English 1–2 staff and most emphatically did not appear in the classroom. This was an English course and not a course in philosophy, though, like everything else worth talking about, it had its epistemological biases. Perhaps the staff could be faulted for a certain disingenuousness in making so little of grander philosophical relations. But no, it was an English course, and I think we were so caught up in our concern for our students' expression of their experience that our place in intellectual history (if any) rarely occurred to us.

One suggestion of Broderick's, however, in that 1958 article, is very worth emphasizing: namely, that English 1–2 was most essentially American. This, too, was not a claim I can recall anyone making at a staff meeting, but it seems to me now exceedingly true. It's hard to imagine another country or another culture where such an enterprise would be likely. Perhaps among other things it was an exercise in patriotism.

V

Baird's own publications, as I have said, are few and fugitive, but they have their devoted admirers—and rightly so. The only book that carries his name on its cover is an anthology of autobiographical passages that he published as a textbook in the early 1930s. (I never heard him mention it.) The introduction's opening sentence will sound familiar: "This book is an attempt to provide materials in the writing of English by directing the student's attention to his own resources of experience." It is a most attractive collection, but Baird obviously found before long other and better ways of getting at the student's own resources of experience.

The articles he has published are listed and briefly described in the selected publications section; here I will add only that, for all their apparent diversity, they can be seen as addressing themes dear to English 1–2. And in almost every case, the articles Baird composed break radical new ground: We have a sense of someone thinking in a fresh way, uncontaminated by the usual academic fudging and triviality. I have space for just one example, and I shall oversimplify. In the article on Defoe that appeared in *The American Scholar* (1958), Baird argues that Defoe invented a new style, one that "relocated the human imagination" by denying "imagination's very existence." That is, Defoe wrote as if he were a combined camera and tape recorder, and "knowledge for the reader becomes direct, unfiltered, firsthand. You are there." In this act of destroying our awareness of any medium of communication, Defoe made it possible for people to tell it like it is, as the current vulgarism has it, without recognizing confusion or the mind's ordering artistry. Think of President Harris. Think of *Time Magazine* or Ronald Reagan. "The victory of Daniel Defoe, that straw-stuffed figure, turns out to be a defeat, and . . . one of the decisive battles of the world has really gone against us."

VI

Baird remains very much a respected and familiar figure in the Amherst community today, though he is probably, for most people, respected at a little distance. In both social and professional interchanges, he has always suffered from the handicap of saying what he thinks. He has no small talk, hates cocktail parties, and couldn't endure committee and department meetings. (English 1–2 staff meetings, of course, were something else.) He complained often about his juniors, and he fought constantly with college administrators. "This is the worst year I ever saw at Amherst: all the fools are now in command"—that remark might have done duty for almost any year. He frightened some people, including those in command. Yet what was always available, if you gave him half a chance, was the most generous and sensitive awareness of others and their feelings. Caring so deeply about the course, he cared deeply about all those who worked with him. Here is a note he wrote to a young and anxious teacher who had finally got to the point where he could compose a halfway decent assignment for English 1:

> Do I ever say with sufficient emphasis to carry how much pleasure and admiration I feel when I see you at work as a teacher. We do take each other for granted and expect as a matter of course the understanding from others—at least in this course—which was hardly come by for the individual himself. I feel I have been especially dull lately and have taken everything for granted. Let me make acknowledgment now and in a loud voice.

At work as a teacher—that is the *key term*. It is his own work as a teacher we acknowledge now—and in a loud voice.

Selected Publications of Theodore Baird

The First Years: Selections from Autobiography. New York: Farrar, 1931; rev.
ed. 1935.
> Anthology for courses in writing.

"The Time-Scheme of *Tristram Shandy* and a Source." *PMLA,* 51 (1936), pp.
803–820.
> Sterne constructed a "coherent and elaborate time-scheme" more or
> less accurately from a specific historical source, so that his book is prop-
> erly seen not as a wildly eccentric structure but as an "exactly executed
> historical novel."

"English 1 C." *Amherst Graduates' Quarterly,* August 1939.
> On early experience with team-teaching.

"Darwin and the Tangled Bank." *The American Scholar,* 15 (1946), pp. 477–486.
> Darwin's great phrases, the struggle for existence and survival of the
> fittest, were figures of speech, and Darwin knew it. Though grounded in
> the most painstaking scientific observation, the *Origin* is also a literary
> achievement, its author very aware of the difficulties of language and
> the limited application of his metaphors.

"Sympathy: The Broken Mirror." *American Scientist,* 37 (Spring 1949).
> "I do not understand my neighbor. I cannot put myself in his place."
> The faculty of sympathy cannot penetrate the mystery of human con-
> duct.

"English 1–2: The Freshman English Course at Amherst." *Amherst Alumni News,*
May 1952.
> Baird's fullest published account of the course. Included is a photo-
> graph of the staff at that time, looking very solemn.

"The World Turned Upside Down." *The American Scholar,* 27 (1958), pp. 215–
223.
> On Defoe transforming language into "truth."

"A Dry and Thirsty Land." In *Essays on Amherst's History.* Amherst: Vista Trust,
1978, pp. 78–138.
> An amusing, heartrending, thoroughly researched account of the con-
> fusions and contentiousness surrounding the founding of Amherst Col-
> lege in the 1830s.

RICHARD BRADDOCK

by Richard Lloyd-Jones

Richard Braddock is the prototype of the professional teacher, scholar, and administrator devoted to the teaching of writing. A practical American, gentle, good-naturedly skeptical and open, he taught and led by his character and thus serves also to exemplify the classical rhetor, the good man seeking truth and community. The blend of character, professional activities, and scholarship is what made his influence on the profession so great and still has an effect even though the body of his published work is quite small.

Braddock was born June 14, 1920, in Glen Ridge, New Jersey. He was graduated from Montclair State College in 1942 and promptly joined the army, serving first as a payroll clerk (staff sergeant) and later as an aerial navigator (second lieutenant). For six years after the war he served in the Air Force Reserve, and he used the GI Bill to resume academic work. He completed his M.A. in American literature at Columbia in 1947 and started his doctorate in the teaching of college English at Teachers College. He received his Ed.D. in 1956. Braddock was a product of the world driven by World War II.

He began teaching before his degrees were completed. He taught at Tenafly High School in New Jersey in 1947–48 and then moved to Iowa State Teachers College (now the University of Northern Iowa) as an instructor in the Department of English and Speech. In 1953 he was made an assistant professor, and after seven years in Cedar Falls he shifted in 1955 to the University of Iowa to supervise the writing component of the Communication Skills Program. He remained an assistant professor of communication skills until 1963, when he was made coordinator of the whole program and was given a joint appointment as an associate professor of English and rhetoric. He was quickly promoted to full professor in 1965. Although the sudden spurt of recognition may have come as a result of steady professional activity and the publication of his major work, *Research in Written Composition* (1963), it also may have been affected by the fact that John Gerber had been made chairperson of the Department of English and recommended that Braddock have the joint appointment. Although the freshman Communication Skills Program was not a part of the Department of English at Iowa, Gerber had at one time been its coordinator and understood the kinds of work carried on by people in the program. Indeed, Gerber was a supporter and interpreter of all kinds of teaching of writing at Iowa, so his support greatly aided Braddock by providing scope for work in composition. It is worth remembering that in the 1950s and 1960s almost anyone doing work in the teaching of composition needed a sponsor from the more prestigious areas of the institution.

To recall Braddock the teacher a decade after his death is not easy even though it is crucial if one is to understand his effect on the profession. One can note the courses he taught: composition at all levels, including Science Writing and Writing for Social Action, English Teaching Methods, Mass Communication, Speech, Literature for Adolescents, a series of summer courses for high school teachers, and various in-service training courses for teaching assistants. He rarely taught literature and did not have a chance to teach much about the research methods he fostered. He always had superior responses from student surveys of opinion, and at his death there were numerous testimonials to how individual students had found their lives changed by having met him. Clearly, he was successful in a wide variety of writing courses.

One can make more useful inferences from his essays on teaching and from materials he prepared for teachers. A brief essay, "Crucial Issues," which appeared in the October 1965 number of *College Composition and Communication,* provides an important insight into why Braddock was an important teacher in the profession. His pragmatic positivism was undergirded by a strong ethical concern. Essentially, Braddock asks that students analyze the "readers" preconceptions and use this analysis in structuring" a composition. In particular, he posits the classroom audience. Braddock says that the student knows this audience, will read the paper aloud to these people, and get their immediate response. Writing and thinking are social acts.

Braddock understands the composition class in terms of coaching writers or perhaps in terms of a large editorial conference. At Iowa the term *workshop* suggests the practice. Although Braddock may have adopted the practice on the basis of what he learned of John Dewey at Columbia or simply from his own attitudes toward people, the cooperative development of writers grew from the literary clubs of the late

nineteenth century. Members (with help from a faculty adviser) read
their poems, stories, and essays to a group and received reactions. In
1900 or so these societies were co-opted into the basic curriculum. The
more literary composition, "creative writing," eventually became at
Iowa the well-known "Writers' Workshop," but the technique often was
used in basic composition courses and was later called "workshopping."
Many graduate students from Iowa's workshop taught in the program
Braddock supervised, and their inclinations were to copy the practices
they were experiencing, so the interests of the assistants undoubtedly
reinforced Braddock's beliefs. Furthermore, Braddock's own classroom
practices seem primarily confined to individual coaching and small-
group interactions.

Braddock's concern with the audience is not just a technique or an
effort at cynical persuasion. He notes in "Crucial Issues" that if students are to make sense of the "human mess," they must "analyze it, see
which propositions are most reasonable or most common" among the
people in the class, and then identify the issues and weight the "more
important evidence of those who differ." The student in reaching out
to understand others tends to reach new understandings. "This process
of understanding others and modifying oneself is fundamental to liberal education. We cannot neglect it because it is difficult. On the
contrary we who teach rhetoric have a unique responsibility here, for it
is we who are granted a portion of the college curriculum to help
students become conscious of what is involved in thinking deliberately,
and, as they develop their thinking in their writing, to help them shape
their ideas into clearer and more responsible communication" (p. 169).
Because these assumptions underlie the best of classical rhetoric, it is
not startling that Braddock urged the change of the name of the freshman program he directed from "Communication Skills" to "Rhetoric."

The particular technique he suggests for using the audience to guide
thought is for the writer to identify the "crucial" issues. Implicitly,
Braddock is concerned with persuasion. Even explanation has to be
understood in terms of persuading an audience to accept an interpretation. Theories of language or rhetoric or literature did not much interest him, but his practice reveals that he was interested in the social basis
of language rather than its formal or representative power. The social
limits of a particular situation determine what Braddock means by
"crucial." He does not mean "big" issues, such as the survival of Western civilization or a national policy on drugs. He does not mean the
stock issues of debate—need, practicality, and the rest. A crucial issue
may also be a major one and may fit the definition of a stock issue, but
Braddock identifies a crucial issue as one that a writer believes is truly
important to the audience, one the reader will find decisive. Braddock
then explains his points with examples, the chief one dealing with

lowering the voting age, but his point is that by thinking carefully enough about the concerns of the reader in a situation where the reader's response is a real one, the writer is forced to gather evidence carefully and concentrate on what is imagined to be the vulnerable point in an argument. One is obliged to consider the audience primarily as a basis for examining the ideas. This is ethical argument in the moral sense although it may be reinforced by ethical proof in the Aristotelian sense of a writer's personal attractiveness and trustworthiness.

Braddock's concern with what was truly within the students' competence is also evident. In most matters he preferred the here and now, the ideas that can be examined with care. He gave examples from the newspaper or campus life. His own language was plain and direct. His advice was practical, as in his several essays and booklets on the dictionary, vocabulary, and spelling. His questions were, "How do you know it? What difference does it make?" Just as abstract speculation bored him, he encouraged students to stick with what they could know. He liked firsthand observations selected so as to make something happen, and he wanted students to succeed. Unconventional manuscripts, sloppiness, and excessive ornament distracted readers, so he helped with mechanics and was sometimes impatient with literary pretension. He taught writing primarily for those who would not major in English but who would do the world's work, but it is also worth noting that he tried to see with the students' eyes in order to meet their own sense of need. That was crucial to him.

He carried this commonsense approach over into his administrative role. While he ran the freshman program at Iowa, two nationwide trends showed up in local dress. The most serious was Iowa's version of the retreat from freshman English that marked so much of higher education after the Second World War. In many schools, the requirements and even the courses were dropped. In others, the courses were kept in name, but very little class effort was devoted to teaching either writing or speaking. At Iowa the courses were kept, and the classroom activities were directly tied to problems of discourse—Braddock, Gerber, the liberal arts dean Dewey Stuit, and others saw to that. But during the 1960s the senior faculties of English and speech withdrew from the courses in order to teach advanced literature or communications courses and devote more time to scholarship. Then the junior faculty found other tasks, too. At first, Braddock tried to hold regular faculty members for teaching honors sections, but eventually he accepted almost total separation of the freshman courses from English and speech. Instead, he developed an administrative structure to compensate for separation by providing suitable participation of graduate students, mostly doctoral candidates in literature or professional writers seeking Master of Fine Arts degrees.

In fairness, we should recall that colleges all over the United States grew very rapidly in the 1960s, and the demands on the faculty far exceeded what could be done. Naturally, faculty members chose the tasks they found congenial as well as necessary. From 1945 to 1946 many colleges doubled their enrollment. The shock forced faculties to consider new ways of working, but the sustained growth of the 1960s, combined with a broadening of the social base of higher education, led to permanent changes in composition programs. In fact, it made the term *program* meaningful. Braddock had a few kindred souls who considered teaching composition the basis of a career. These faculty members all taught some classes of freshman, but mostly they taught graduate assistants and shared the administrative chores of the program. In short, Braddock made the best of a difficult situation and, in doing so, created a powerful educational tool. One version of the plan he reported in *College English* (in October 1970) under the title "Reversing the Peter Principle to Help Inexperienced Graduate Assistants Teach Freshman Rhetoric."

The details of the plan are less important than the general principles. In the 1950s, much of the talk about administration in universities had been devoted to economies of scale. As mass production had been a major American contribution to industry, so mass education would reduce the costs of running colleges. Those who defined education primarily as transmission of knowledge and who were most comfortable with lecture courses had no trouble with the implications. They depended on large staff meetings and standard syllabi, which sometimes offered a day-by-day list of materials to be covered. Much of our present enthusiasm for mass testing was nurtured in this era. Those teaching the classes were clearly underlings. In practice, this industrial model often broke down, and the hired hands did what they pleased, and no one noticed, but the social differentiation was real.

Braddock was by temperament egalitarian, and Iowa had a tradition of easy consultation and relative indifference to rank. The tone of the program to supervise graduate assistants was supportive. The graduate assistants were to be treated as younger colleagues. Those who were particularly apt were in subsequent years chosen to assist with the training so that every session could be jointly led by a permanent staff member and a graduate assistant. Either or both might lead a particular seminar; the new assistants were free to get help from either. All assistants were partners in the operating decisions relating to the freshmen and to their own work as teachers. As graduate students, they were members of a quite different departmental structure. Especially after the mid-1960s, their graduate professors were hardly aware of the Rhetoric Program; warm, personal relationships directed toward literature did not require discussion of composition.

Even the process of selecting teachers had two steps: the graduate department provided a list of those eligible for appointment, but the Rhetoric Program chose its own teachers from the list. The in-service training was neither didactic nor dictatorial. Except for a few meetings of assistants as a group to create a sense of faculty, the work was done in small groups, following a plan of discussion set up by the seminar leaders in consultation. The system was one of self-government within a limited context—practical and open-minded. In an ideal world, perhaps composition and literature would not have been so much separated, but the system did make certain that composition actually was taught. So also in an ideal academic world the graduate and undergraduate faculties would not be so much separated, but the pragmatic Braddock was more interested in getting the work done than in holding out for the ideal.

Braddock's adjustment to the problems of having a transient staff is a pilot study for many current blends of permanent and temporary faculties. Even though the problems of the multisection course provided a stimulus to formation of the Conference on College Composition and Communication (CCCC) in 1949, the heavy dependence on short-term teachers of composition in colleges of all kinds is a commonplace of the 1970s. Braddock's model of administration, a function of his social values, is the polar contrast with that of the industrial model.

Temperament as well as social values made the system work. Braddock's ability to work with the people of the 1960s is an indicator of his character, and it forestalled many revolts against authority as such. The Civil Rights Movement and later the Vietnam protests provided a background of turmoil; soothing the pricks of liberal consciences seemed to require grand gestures designed to make administrative life difficult, and indeed some academics seemed unable to respond to conscience without a bit of pressure. Also most schools were still unevenly adjusting to increases of enrollment that changed the general relationship between faculty and students, between general and professional education, between scholarship and teaching. The old order was finding change very difficult. The graduate assistants were especially vulnerable in these unstable situations, and an administrator whose faculty was almost entirely made up of graduate assistants faced an almost impossible task. Braddock responded to iconoclastic gestures with sympathy, patience, a sense of humor, and a clear sense of obligations to freshmen.

The administrative structure that encouraged a joint front of faculty and graduate students helped, and most of the graduate assistants cared deeply about their students, but Braddock's genuine interest in the views of other people was even more important to maintaining the basic work of the Program. A trivial example suggests the tone. An extremely bright doctoral candidate whose pacifist views reinforced a

deep distrust of all administrative machinery was given to strolling on a
fourth-floo: parapet. The perch was perilous, and his casualness
caused great administrative alarm—to no effect. After all, this assistant
was a Quixote, who on another occasion commandeered a motorized
lawn roller as Rosinante and went off to challenge the bookstore profi-
teers. Braddock did not shout orders. He merely reminded the man of
the accident forms that would have to be filled out if there were to be a
fall. The crucial issue for this man was bureaucracy, not safety, and
Braddock was shrewd enough to notice. He was very good at imagining
values other than his own and then finding ground on which dispu-
tants could stand together.

Notable though Braddock's local achievements were, they represent
the character rather than explain his impact on the teaching of writing
in the nation. His "middle-class ethics" enhanced by a Deweyite education forced him to make his beliefs active in the professional marketplace. Democracy is not passive. He could not be merely an observer
however much he valued systematic observation. At all stages of his
career, he was a leader of those who led. He was a teacher of teachers.
Helping people think wisely together was an important goal. That, too,
was a product of character.

Braddock joined the National Council of Teachers of English
(NCTE) as soon as he started teaching in 1947, and he joined the Iowa
Colleges Conference on English (ICCE) as soon as he came to Iowa and
was its president 1958–60. He was a founding member of the Iowa
Council of Teachers of English (ICTE) in 1955 and acted as its president during 1957–58 and led ICCE to assume a second identity as the
College Section of ICTE. He would later receive its Distinguished Service Award. By 1959 he was already arguing for increased standards
for teachers of English by proposing a "bar exam" as a means of establishing professional status although he also worked energetically to increase requirements for certification in more conventional patterns. He
served with both the Iowa Board of Regents and the Iowa Department
of Public Instruction to coordinate education programs and improve
the professional status of teachers.

To enhance the standing of composition teaching in the high schools,
Braddock arranged for Iowa colleges and the larger school districts to
form a program to give college credit for specially designed high school
writing courses. In part, he was responding to pressure to give freshman credit for advanced placement (AP) courses by creating a suitable
alternative. He perceived the AP work as being about literature and
thus not suitable for credit in place of a writing course. More important
than the actual design of his Advanced Standing Program (ASP) was
his procedure in establishing it. College people visited high school
classes. High school people visited college programs. All talked to-

gether about how the work was to be done. The major universities agreed to establish summer workshops to prepare and certify the teachers who would offer the courses. It was a highly successful political effort of consensus making to improve the curriculum and enrich the community of teachers by improving some of its leaders.

Another dimension of his professional efforts to bring high schools and colleges together was directing four summer programs for teachers as well as serving as a visiting faculty member in the summer at Harvard and the University of Oregon. The best-known of the summer programs were institutes created under the National Defense Education Act (NDEA). The ones at Iowa followed the general pattern of the tripod metaphor embodied in summer programs designed in 1960 by the College Entrance Examination Board. There were courses in composition, literature, and linguistics as well as a common period for the discussion of curricula. Students were practicing teachers who were selected and financed under terms of the federal grant. Braddock drew his institute staff from colleagues not especially concerned with high school teaching, and he provided the energy and commitment to make powerful programs as well as courses. Sometimes he worked with those already committed, like Carl Klaus, with whom he team-taught a writing course. Klaus, who had come to writing through a conventional literary degree, later directed several institutes and became much engrossed in problems of education in the schools. Klaus led Braddock to appreciate the arts of literary prose more, and Braddock led Klaus to a greater sympathy for the mundane prose of daily business. The two of them offered a demanding and exciting course for teachers. Klaus later organized the Iowa-NEH Institute on Writing for directors of college freshman composition programs. The fusion of literary and rhetorical training was strong in Klaus's institute, and if Braddock had lived to share in it, he might well have accepted more of Klaus's concern for writing as art.

Still, Braddock's greatest influence came through his professional activities in the NCTE and CCCC. They enlarged his work in Iowa. He was a member of the committee on the preparation and certification of teachers and was associate chairperson of the committee dealing with the publications of affiliates. He chaired the Iowa part of the NCTE Achievement Awards program, and he served on both the Resolutions Committee (1960) and the Nominating Committee (1961). His main contributions to NCTE dealt with research: cochairperson of the committee reviewing the state of knowledge about composition (1961–63); chairperson of the standing committee on research; trustee of the NCTE Research Foundation; and founding editor of *Research in the Teaching of English*. From these efforts, in turn, grew his most noteworthy publications.

Braddock worked in CCCC during the period it consolidated its functions. He joined the conference in 1952 just after it had absorbed the initial thrust of dealing with the overwhelming freshman enrollments following World War II. The first members were driven by their immediate needs, but the crucial period for any group comes when the initial pressing needs have been met and the organization must develop its permanent identity. It was in the second decade that Braddock served on the Executive Committee of CCCC. He then became secretary and finally entered the three-year electoral sequence culminating with his being chairperson in 1967. He pushed the group toward a broader base for its membership, including community colleges, toward what would now be called affirmative action, and toward scholarship. At Iowa after careful consultation with two-year college teachers and administrators, he developed a special Master of Arts/Education Specialist Program for the training of community college teachers, and he helped CCCC with the committee work that led to the creation of what are now the two-year college regional organizations within CCCC. He probably did not foresee large regional conventions, but he accepted the principle that higher education should be accessible to all, and he believed that the community college was the most suitable structure for serving much of the population. His goal in developing the M.A./Ed.S. was ensuring professional competence in areas where most two-year college teachers would spend that time.

Although CCCC had from the start been relatively sympathetic to the "protected" groups of affirmative action, perhaps because composition courses often were taught by faculty members excluded from the mainstream and, in the 1960s at least, taken by students perceived as outsiders or as deficient, still Braddock pushed the issues of representation by his appointments and policies in the organization. His sympathies were with the underdog; he was driven to draw talent from those who did not believe in themselves. Quite possibly, part of his interest in the development of the two-year colleges was that he understood them as serving people who had been neglected. The important effect of his efforts was the result of many small decisions rather than of any grand gesture, but it can be illustrated by the events of the CCCC convention in progress in Minneapolis at the moment that Martin Luther King was murdered. Braddock simply threw out the published convention program and turned the meeting over to the concerns of blacks in the society. That was, in fact, the crucial question for all Americans on that day—and for teachers especially—but it would have been easy to pretend that it was too political for an academic meeting. Perhaps all good composition teachers are driven more by ethical concerns than by purely intellectual ones, and Braddock was especially concerned with the common good. He felt that those who most nearly represented the outraged in society must have the audience on that day. But that decision was merely a more public version of characteristic choices he made.

The scholarly work that for most people anchors Braddock's career is *Research in Written Composition* (*RWC*), published by NCTE in 1963. Along with Joseph Miller, Braddock had organized and cochaired a council committee charged with the responsibility of discovering the state of knowledge about the teaching of composition. Such concerns had been aroused by discussions at the Basic Issues conference of 1958, a joint effort of NCTE, MLA, and CEA to redefine the goals of teachers of English. By 1961, NCTE (in continued association with the other groups) published *The National Interest and the Teaching of English,* and the U.S. Office of Education had been persuaded to create Project English as an agency of the Cooperative Research Division. With Braddock's prodding, the committee surveying the state of knowledge about composition compiled a list of some 500 research essays and monographs. It was evident that a large group could not complete the charge to the committee—to report the state of knowledge. Braddock, therefore, recruited two colleagues at the University of Iowa and, with financial help from the University and the U.S. Office of Education, began obtaining, reading, and evaluating these published studies. Lowell Schoer of the College of Education was to cover research design and procedures; I was to supply the perspective of one interested in discourse theory; Braddock as the organizer was the unquestioned leader of the trio.

Our work took a year and a half. Many of the studies were available only on microfilm, which was slow in coming, and we all had other responsibilities, but we met two or three afternoons a week to compare the results of our week's reading. Early on, we found that the mountain of material had to be reduced by definition, for we kept adding new titles. The original committee had not found every suitable study. We ruled out oral composition, pure theory, surveys, handwriting, and anything that did not specifically deal with writing. Perhaps the omission we felt most keenly was Albert Kitzhaber's *Themes, Theories, and Therapies,* for we felt it was an important survey of practices in college teaching and should have alarmed us all, but Braddock was properly firm in holding us to our limits. As usual, he argued for keeping the task to a size that could be done thoroughly even if that meant putting aside important issues. He wanted the book to be the last word within its own limits and did not want a reader's attention to be scattered. Almost universal citation of the book suggests that he was right in setting the limits even though some research was still overlooked.

Probably the most significant use of the book has been as a guide to research method—or perhaps to presenting the results of research. Often the studies we examined seemed to lack controls although perhaps sometimes the reports were just incomplete. Braddock's predilection toward exact observation and fairness in comparisons made us very fussy about each step in comparative studies. The rules for inclusion of research in our report at all favored those studies with comparisons between experimental and control groups. Braddock was a thorough empiricist. Schoer often argued for the validity of other kinds of knowledge, most commonly "received lore," but we all insisted on precise observation reported in detail. I was the one most likely to argue that a particular hypothesis was not worth the bother to explore because it was narrow or limited in application or merely cosmetic, but we regularly found that the broader the hypothesis, the greater the likelihood of incomplete reporting or unfair comparisons. In the end, almost no study could survive such skepticism unscathed even though we were convinced that many of the studies dealt with sound propositions. We believed in the soundness on grounds other than those actually reported in the study, however.

The section of frequency counts especially pleased Braddock, and his observations guided the preparation of his later essay on topic sentences, which won the first Braddock award established in his honor by CCCC. Essentially, one identifies an observable feature in a written passage and counts how often it appears. In the *RWC* passage, Braddock noted the difficulty of defining countable categories, so he called for numerous examples to make abstractions concrete. He also observed that different researchers examining the "same" problems used different systems of classification and that made comparison impossible. His examples of noncomparable categories dealt with error counts—the enumerating of nonstandard forms in passages of writing, and he lamented that no standard system then existed. He remained optimistic that, with the help of linguistic studies then in progress, some system would be developed. Subsequent error counts have not been standardized, as a look at error counts reported by the National Assessment of Educational Progress (NAEP) will confirm, because the researchers do not share a theory of language and syntax nor even a common view of error. An acceptable theory will have to precede the formation of standards, and few now would dare to have Braddock's optimism.

Braddock also objected to lists of errors that concealed the importance of differences between items on the count. How many "errors" of what kind were significant according to what principle of significance? He probably undervalued the usefulness of reporting groups of items according to an articulated theory, but he was right in rejecting raw numbers or simple lists and in asking that counts be interpreted by use of percentages of occurrence in so many words of text. He also sought counts of complex rhetorical structures and situations, that is, counts of features not representing errors, for he felt that description should be separated from a tendency to prescribe. Because he recognized that the number of possible items subject to counting is extremely large, he argued that researchers should find sharply limited features that correlated well with larger ones. He did not discuss the problems of demonstrating the relationship between the particular items chosen and some more inclusive idea, but apparently he thought it could be argued by other means. For example, does the number of adjectives in a passage correlate to exactness of definition?

When he prepared his widely cited study of topic sentences, he followed his own advice. He wanted to consider the accuracy of generalizations about the proper use of topic sentences. That is, he chose a larger rhetorical situation not requiring a concept of error and depended on the definition of a particular discourse form, which he illustrated. He chose as his sample writing from the world of affairs, not from school, and he reported the details of his sample carefully. These were twenty-five essays from substantial popular periodicals. He numbered each paragraph and marked off the T-units in each sentence and finally identified which T-unit was the topic sentence. What he found was a limit on schoolbook advice about topic sentences, but he could not, on the basis of his sample, propose a comprehensive substitute for the rules. The finding is probably less important than the care that he illustrated in his procedure. In the end, he offered to make his raw materials available to other scholars who might want to check his work, but he died before the article appeared, and the materials were lost.

A person who had been challenged by Braddock's observations probably could rethink the basic assumptions in order to rationalize what he described. Those who have read and written much as well as those who have listened and spoken develop a tacit knowledge of how to guide a reader through the structure of a discourse, but the range of variation is so great that no simple rule applies. Situations are sufficiently unique to make their own patterns. We have become used to the linguists' assertion that the producing of each sentence is a creative act, so we ought to expect even more in longer discourse. Textbook writers feel that pressure to produce rules, and the rules may even be useful although falsely overstated, but the overstatement may also lead to frustrating misunderstanding, as Braddock showed. Teachers begin to claim too much and thus give bad advice. Braddock's article becomes an excellent demonstration of carefully controlled observation that make one skeptical of easy answers; it also satisfied his own desire to illustrate some of his prescriptions about method in research studies. He defined exactly, illustrated his definitions, was explicit in reporting his methods and results, and wrote plainly.

Other sections on method discussed in *RWC* are less distinctive although still useful. Braddock's advice on rating compositions is mostly directed toward controlling variables—or at least recognizing variables in reporting results. Although considerable refinement in rating methods has been made in the twenty years following (cf. Cooper and Odell, *Evaluating Writing*, NCTE, 1977), his effort to name the variables affecting the writer, the assignment, the individual rater, and the collective rater is still the basis for organizing critiques on the scoring of essays. Similarly, another feature of the book presents what is now standard advice. Braddock emphasized the design of projects and pilot studies. Others have provided advice more elaborate than his, but still well within the framework he proposed, for he clearly wanted observation controlled by careful planning. Adjustments to a study after it was done did not please him.

The original purpose of *RWC* was to report the state of knowledge about composition. The brief answer—very little—is misleading because the qualifier to the term "knowledge" is often ignored. The study dealt with empirical research based on actual samples of writing. Other kinds of knowledge exist. The experience of teaching the skills of writing during five millennia has supplied a large body of lore well tested and quietly adjusted through empirical observation. The need for another kind of knowledge was created by the astonishing increase in the size of schools in this country after World War II. High school diplomas became a standard expectation. Mass instruction and scientific enthusiasm seemed to require explicit rules. The real question for the committee was whether the empirical studies of the first two-thirds of this century had provided useful insights for mass education. The answer of the book seems to be mostly that they had not, but that some useful lines of questioning had been established.

The reported state of knowledge is dreary reading. Studies of the nature of the student writer and of the psychological processes that generate writing were barely begun. Most were surveys. Major cognitive studies of the writing process have come since—and probably in response to suggestions from sources other than *RWC*. Perhaps the most significant note is the call for more longitudinal studies, like that of Walter Loban, which was well started by then but far from finished. The most popular subject for study was instructional method. The questions seem to invite immediate practical response, and that may account for the relative ease with which studies were funded as dissertations. Institutions may even have pressured graduate students to have the work done. Often such studies were too much limited in context or pushed a particular instructional device and thus offered little general knowledge. Often articles that did not fit the *RWC* definition of "research" but rather described and rationalized practical proposals for teaching were more persuasive in presenting "knowledge."

One of the most frustrating areas of research dealt with the role of tests, for these are important in administration of writing programs and in research. They are by-products of size, for no one can know even a substantial minority of the students. Administrators want assurance that all members of the mass are being judged on a single scale. Although much work has been done since 1963. I think the advice of *RWC* still applies because the basic conditions haven't changed. Although these bits of advice were based on a number of studies, they also represent the pragmatic administrative experience of Braddock. *RWC* notes that mass testing is at best approximate in reporting on individuals and often misleading in aggregate reports on groups. For placement in courses, objective tests may still suffice and are relatively cheap, but administrators then must allow for easy appeal from the decision of the test and should assume that the test gives little indication of why a student is placed in a particular course. If one wants to know more specifically what kind of instruction is needed, then one must have writing samples. Because information about the writing itself is useful primarily to the classroom teacher, it is best to have the papers read by the teachers, and inexperienced teachers can get a kind of in-service training by reading papers in the company of those who are experienced. Final evaluation of student writing, apart from a course or in situations granting credit, probably requires multiple readings of more than one sample of writing. This is still standard advice.

Although a number of other specific questions were discussed in *RWC*, only one has received enough continuing comment to justify mentioning it. The large number of studies about "grammar" and composition seemed to require comment even though most were unsatisfactory as studies. They rarely were based on similar systems of classification and often seemed not to deal with what a serious linguist would call "grammar." They preceded the developments of generative grammars and often were confused in mixing traditional and structural grammars. They were conducted for much too brief periods. Yet it was hard not to read them without feeling that the abstract knowledge of grammar had little to do with the experiential process of producing texts of writing. Even questions of whether the knowledge of grammatical terms helped in revision or in discussing writing were unanswered. Given excellent writing produced by people who were ignorant of grammar, the whole issue of "grammar" seemed beside the point and thus likely to steal time from useful practice and response. But even though the studies were not comparable and were questionable as examples of method, the observed details all pointed in the same direction. School grammar hindered writers more than it helped.

Of course, that did not settle the question then or now. The study of language is valuable in itself. Braddock himself prepared a beginner's anthology of essays on linguistics. *Introductory Readings on the English Language* appeared at the same time as *Research in Written Composition.* Partly he was responding to a fashion for "controlled research papers." Collections of readings on single topics (American architecture, the writings of Steinbeck, or world government, for example) relieved the load on understaffed and understocked libraries and challenged students to form propositions based on multiple sources. Mostly, though, Braddock believed that knowledge about language was "a highly significant area of . . . liberal education." To be sure, he also added that he hoped the book would alert students to ways they could communicate with clarity, felicity, and responsibility. The essays included in *Introductory Readings on the English Language* illustrate the first goal of providing general knowledge, however. Braddock did not claim to be a linguist, but he did recognize public controversy. He dealt with the history of language; with dictionaries; with technical accounts of grammar and popular accounts of usage; with punctuation and spelling; and with general semantics, logic, style, and rhetoric. The essays are good reading even now, and Braddock's contribution was that of a teacher of beginners, but the essays also illustrate that he found language fascinating, useful for freshmen. He did not believe that such study was particularly relevant to the writer in any specific way even though it might help the teacher of writing, but he did believe that an understanding of language in general terms could free a writer to use the language more vigorously.

He was realist enough to know that many people who call for teaching grammar really mean to ask for the teaching of usage acceptable to a particular group. By definition, usage records the practice of speakers and writers, and disputed usages arise when subgroups of language users make different choices. One form in itself is not superior, but the people who use it may be more powerful. The concerns of people who prescribe usage are thus political rather than intellectual. Still, the section of the anthology on usage is instructive by what it doesn't include as well as what is provided, for Braddock did not make this understanding explicit. Almost all the space is devoted to an excellent comprehensive statement by Porter G. Perrin from the third edition of *The Writer's Guide.* Short essays by Wilson Follett and Bergen Evans are included to stimulate arguments, and yet nothing is included in anticipation of battles later in the decade. In *RWC*, Braddock's statements about the effect of instruction in grammar, about "error counts," and about objective tests (which are often tests of usage) did not address the political issues of usage as such even though these are deeply embedded in the social effects of language, a crucial part of Braddock's later concerns. He sympathized with the policies later expressed in the CCCC statement on students' right to their language, yet he was more concerned at this time about issues of structure and evidence, about fairness and civility. He so much preferred the plain and common style that he didn't pay much attention to surface variation that sometimes leads to elegance, sometimes to obscurantism. He didn't want writing to call attention to itself. In any event, he did not pay much attention to usage and dialect as evidence of social or ethnic class, perhaps because Iowa provides such a homogenous student body that he didn't need to.

RWC became a rallying point for those concerned with the teaching of writing in schools. Perhaps because popular knowledge about language was often false knowledge and popular prescriptions for the schools seemed simplistic, perhaps because faith in empirical studies based on the social sciences was high, perhaps because many teachers were unhappy about what was being done to teach writing, perhaps because the federal government seemed likely to offer money, many called for additional research. If only we knew more, we'd do better, they seemed to say; and in the lore from our ancestors, truth and falsity are mixed. Research will tell us what to reject. Braddock himself saw the pre-1960 research as akin to alchemy. He called for new studies that would have the authority of chemistry. The call for "new knowledge" came at the same time that "new" students were pouring into the two-year colleges and needed extra instruction in writing. The Civil Rights Movement and the Vietnam War were challenging voices of campus authority, so faith in the old means of instruction probably needed bolstering by scientific method. Into this context Braddock placed *Research in the Teaching of English* (*RTE*), a new journal first published by NCTE in 1967 after a year or so of careful planning. Although it was a new kind of venture for NCTE, it was a suitable follow-up to *RWC*, which NCTE had published.

RTE was for Braddock a natural response to the survey of knowledge. He wanted a practical contribution to a real solution of the problem. If the present research is poor, provide a respectable and predictable place of publication for new studies, one that might encourage research in a style he approved and might earn prestige for the researcher. The first issue had remarkably little by Braddock himself— brief headnotes and some news items. The style reflects his background in educational research and empirical methods. Two of the six articles were general comments on research methods. The other four articles included a count of features in prose ("Sentence Structure and Prose Quality"), a study of teaching methods, a description of the background of poor writers, and a study of correction techniques, all subjects raised in the *RWC*. The largest "department" included research abstracts and a bibliography by Associate Editor Nathan Blount; there was also a Roundtable Review of *The Measurement of Writing Ability* by Godshalk, Swineford, and Coffman.

Braddock continued editing *RTE* through 1972. By the second year two articles on literature crept in—one each on teaching poetry and on teaching short stories. Eventually, the teaching of literature took up a fourth to a third of the space in the journal, but Braddock clearly preferred composition. He quit writing the headnotes; his consulting editors did the previews of articles they recommended. Blount continued to do the bibliography during Braddock's six years. The self-effacing editor was revealed only in the choice of articles and in setting topics for the Roundtable Reviews. These usually dealt with composition; the last topic covered the first National Assessment of Educational Progress report on writing, as reviewed by John Mellon and Sr. Mary Philippa Coogan with a response from Henry Slotnick of the NAEP staff; that discussion led to major changes in the assessment and subsequent inclusion of "writing people" along with the professional testers.

The importance of Braddock's editorial creation transcends any set of articles. *RTE* has passed through other editorial hands and is identified with NCTE, not with an editor. In its own way, it made explicit a council commitment to research as well as to teaching. In some ways, its narrow limits may have shifted English too far toward the social sciences. For Braddock, it was a way to follow up the implications of *RWC*, for one result of the survey of the state of knowledge was to put great stress on the empirical sources of knowledge about composition. Braddock was not a theorist; he did his work within a range that might be paralleled with Thomas Kuhn's term "normal science." He was practical, responsible, industrious, and open-minded; and he set these qualities into the framework of the journal. But although he was liberally educated and humane, he probably was not a humanist.

The professional achievement of Richard Braddock is in a large way reflected by the reaction to his death in Australia while crossing a street. The death was sudden, of course, and it happened to a relatively young person who might have been expected to contribute more to the profession. That always brings crowds to the funeral. Still, the outpouring of regret exceeded the amount that might be generated by sadness about the loss of a colleague. Letters, speeches, formal resolutions and elaborate tributes, and the Braddock Award were among the responses to the whole person as well as to what he had done to shape the profession. He gave rigor to empirical research in composition, raising the quality while enforcing a kind of limit on the imagination. His leanings toward social science made it easy for him to reach across disciplines and represent the importance of composition. By appealing to conventional academic virtue, Braddock raised the status of all teachers of composition. He gave their role a higher regard in the academic marketplace, but the essential part of his power was that he truly prepared for the future by melding individuals into the community. His heirs in the community have not necessarily followed his intellectual lead, but they often have accepted his sense of profession. No wonder there was mourning for such generosity of spirit.

Selected Publications of Richard Braddock

BOOKS

Introductory Readings on the English Language (ed.) Englewood Cliffs, N.J.: Prentice-Hall, 1962.
 A textbook containing articles on linguistics, semantics, and communication theory.

Research in Written Composition. Urbana, Ill: National Council of Teachers of English, 1963.
 Braddock's most influential book. It is a thorough examination of previous research on writing and includes an enormous bibliography.

A Little Casebook in the Rhetoric of Writing. Englewood Cliffs, N.J.: Prentice-Hall, 1971.
 Textbook for freshman composition.

ARTICLES

"An Introductory Course in Mass Communication." *Journal of Communication,* 6 (Summer 1956), pp. 56–62.

"English Composition." In *Test 3 of the Iowa Tests of Education Development.* Chicago: Science Research Associates, 1960.

"Crucial Issues." *College Composition and Communication,* 16 (October 1965), pp. 165–169.

"Teaching Rhetorical Analysis." In *Rhetoric: Theories for Application,* ed. Robert M. Gorrell. NCTE, 1967, pp. 107–113.

"English Composition." In *Encyclopedia of Educational Research,* 4th ed., 1969, ed. R. L. Ebel, New York: Macmillan, pp. 443–461.

"The Frequency and Placement of Topic Sentences in Expository Prose." *Research in the Teaching of English,* 8 (Winter 1974), pp. 287–302.

MINA SHAUGHNESSY

by Robert Lyons

Unlike other teachers who have made a significant difference in the field of composition, Mina Shaughnessy devoted herself entirely to a special group of students, those who demonstrated such limited skills as writers that their presence in college classrooms became a subject of intense debate. All her work in pedagogy concerned the relationship of the English teacher and the "basic writing" student—a designation now widely accepted because she used it. Shaughnessy identified with and spoke for such students, asserting their right to a place within higher education. She evoked and articulated their aspirations, while unflinchingly recording their deficiencies as writers and systematically creating patterns of instruction to meet those deficiencies.

Shaughnessy undertook the work for which she is remembered relatively late in her life, and she died much too soon, at a time when she had just achieved a national reputation and when her writings had begun to have a profound impact on her profession. That impact, considering her shortened professional life and the limited number of her publications, says a great deal about her special grace and authority both as a writer and as a person. Those who knew her or even met her briefly would not be likely to forget her. As someone who did know her, I still find it difficult to accept her absence and to regard her as a writer and teacher to be appraised rather than solely as a colleague and friend to be mourned. Nevertheless, her public accomplishments are her memorial; they will be remembered for their intrinsic importance and for their broader influence.

By the end of her life she had become the most widely respected authority on basic writing in this country; her views on her subject were

171

being reported not only in academic journals, but also in the pages of *The New Yorker* and *The New York Times*. When her book, *Errors and Expectations*, was published in 1977, it was called "a force that can redirect the energies of an entire profession" by one reviewer,[1] and *The Nation's* book editor said it "may be the most significant advance in years toward what Benjamin DeMott rightly calls 'the grand project of this society, that of democratic realization.' "[2] She was honored with a special citation by the head of the National Endowment for the Humanities, who observed that "her compassion and understanding of human problems have contributed significantly to bring the humanities to thousands who have never been touched by them before."[3] After her death, the Fund for the Improvement of Postsecondary Education created, with the support of the Carnegie Corporation, the Mina Shaughnessy Scholars Program, commemorating her achievements by funding research to provide better learning opportunities for college students. In a field often marked by controversy and division, her work was invariably accorded attention and respect.

Shaughnessy's prestige stemmed from her ability to take on the differing perspectives of both the basic writing student and the academy. On the one hand, realizing that intelligence is not the monopoly of the privileged, she could stand outside the academic gates with these students and challenge what seemed to her the misguided judgments of the gatekeeper. On the other hand, she stood within the walls and acknowledged the legitimacy of the standards of competence that had so frequently been used to bar the gate. Other writers engaged in presenting the basic writer to the academic world in the 1960s and 1970s saw only that world's intractability and hostility. They created extended profiles of disadvantaged students, describing their social and economic circumstances and explaining their mores and values. The thrust of such descriptions or polemics was to argue ways in which the styles and perceptions of these new students, unfamiliar to most teachers, could, when honored, revitalize the college curriculum and reorder its priorities. Shaughnessy concentrated instead on the student as writer and accepted academic discourse as a given; her focus was on how far basic writing students had to go to master this discourse and how one could best help them to get there. Indeed, she was not caught up in political defenses, because she took for granted the student's right to be in the academy and simply moved on to explore how teachers were going to provide them with proper opportunities for what she often called their "entitlement," an academic education.

Shaughnessy's ability to balance the claims of basic writers and the academy had its roots in the shape of her own career. Born and raised in South Dakota as a rancher's daughter, she graduated from Northwestern with a degree in speech. When she came to New York several years later, she was lured by the possibility of becoming an actress (something that helps to explain her extraordinary presence and self-possession as a public speaker). Put off by the insecurities of a life in the theater, she turned to the printed instead of the spoken word. She took an M.A. degree in literature at Columbia, served as a research assistant to Raymond B. Fosdick, the former president of the Rockefeller Foundation, helping him prepare two of his books,[4] and she worked for several years supervising copy editors for the publishing house of McGraw-Hill. When she began teaching, working part-time in the composition program at Hunter College, she was already in her thirties, and her résumé was hardly different from those of many bright young women who come to New York with a college degree and belletristic interests. Similarly, in the beginnings of her teaching career, she was representative of the many women who make up the majority of the part-time writing instructors in the nation's public colleges, constituting an essential, if unrecognized and underpaid, part of the academic community. From 1962 to 1967, Shaughnessy was a member of that floating population of part-time teachers, working at Hunter College and at Hofstra University, until, in the fall of 1967, she went to City College.

The appointment at City College represented Shaughnessy's first full-time teaching job, but it was a significant change in more important ways. She was persuaded to apply to City College by a friend from Hofstra, Alice Trillin, who had become interested in working with minority students. Both women turned down opportunities to teach in the English Department at City College and became members of the college's SEEK Program—a special program for students from low income areas of the city and, therefore, one that enrolled primarily black and Hispanic students. For Shaughnessy, this was new territory; her previous experience had been in conventional introductory literature and freshman composition courses. The SEEK Program was still at an experimental stage and its writing courses in a state of some confusion. When the writing coordinator became seriously ill at the end of the spring, the administrator of SEEK made an extraordinary decision, prompted, it seems likely, by both insight and desperation. He asked Shaughnessy, whom he had just appointed as a teacher for the following fall, to direct the writing program. Thus, suddenly and unexpectedly, in her first days at City College, she had assumed the role that would be the focus of her energies for the rest of her life.

The SEEK Program, begun in 1965, served as a pilot program for open admissions in the City University, and it provided Shaughnessy with her first experience with basic writing students. Writing teachers in the program were hired by the English Department but were assigned exclusively to SEEK courses. Regular faculty members in the

department were never assigned these courses, and those who volun-
teered often felt the department discouraged this kind of teaching. For
Shaughnessy, the experience must have reinforced a sense of being in
the academy, but not of it. That anomalous position had been the mark
of her earlier teaching career as well because she lacked the doctorate
and had lived the precarious professional life of the part-time teacher.
Such experiences made basic writing teachers generally see themselves,
with justification, as a beleaguered lot. Within the teaching community
of the college, they found themselves outsiders, a circumstance that
often intensified the identification they had made with their students,
those other educational outsiders.

In this social context, it was therefore remarkable that Shaughnessy,
though committing herself to these students, never saw that act as a
repudiation of the academic system that kept itself so carefully aloof
from the work she did. It may indeed strike one as surprising that
someone with Shaughnessy's background would find this particular vocation to start with because the disenfranchised urban poor seem a long
way from the rural Midwest of her youth. But perhaps it is more
surprising that, once she became deeply involved with basic writing
students, she maintained her steadfast commitment to traditional standards for literacy. Those who knew her and shared her concern for
basic writers were often irritated by the degree of deference she
showed to the forms of the academy, for she had ample reason to know
that a doctorate gave no guarantee that its possessor was either humane
or lettered. Part of her deference could be traced to the self-doubts
generated by her lack of conventional credentials, but part certainly
reflected a deep respect for what was best in the academic tradition and
a desire to honor it.

Shaughnessy's own Western heritage, furthermore, led her to stress
the need for individual purpose and effort, as well as the legitimate
rights of individuals. Significantly, her comments on basic writers always assume their own desire to control the language that will give
them full access to the academy and greater opportunity in the larger
society beyond. But she also felt such access meant the opportunity to
experience education as she had experienced it; for Shaughnessy, education had fostered, if not created, the love of language and literature
that had so enhanced her own life. Her taste in literature inclined
toward writers of stylistic as well as intellectual complexity: early in her
career she published an essay on Henry James; she had a long-standing
ambition to teach a course on Milton; she particularly valued Bacon
and Arnold as essayists. One of the paradoxes of her work was her
willingness to respond with such attention and sensitivity to some of the
most elaborate stylists among canonical writers and also to writers con-
ventionally defined as the most inept users of the language. In fact, it

was her own responsiveness to the achievements of style that made her capable of recognizing the gestures toward style of the uncertain basic writing student.

Shaughnessy was an administrator, moreover, as well as a teacher. As an administrator, too, she necessarily mediated between the program she ran, with its educational and social imperatives, and the college that, with some hesitation and discomfort, sponsored her program. Her purposes were not shared by many of her colleagues on the faculty, particularly among the senior and tenured professorial staff. Such divisions were greatly intensified when, in 1970, the university adopted a new policy of Open Admission, guaranteeing admission to a college of the City University of New York (CUNY) for any graduate of a New York City high school regardless of the student's high school program or grade average. The primary source of pressure on the university to make this change had come from the city's minority groups and from social activists who found in the disparity between the racial profile of City College and that of the surrounding Harlem community a dramatic symbol of the educational exclusions suffered by the poor and by minorities. So the center of the whirlwind was City College, and the debate there was long and bitter; the normal life of the college was frequently interrupted by public meetings, demonstrations, sit-ins and walkouts, and even sporadic violence. In the fall of 1970, then, basic writing was no longer the concern of a relatively small and isolated "special program"; it had become a focus of attention for virtually every faculty member, as Shaughnessy now took on the task of educating many more poorly prepared students in a writing program primarily staffed by new and untrained teachers and by uncertain or reluctant professors of literature.

The experience was exhilarating, for adherents of the basic writing program were caught up in the sense of a great social and educational experiment; it was contentious, for opponents of the program were insistent and their antagonism was deep-rooted. Most of all, it proved to be frustrating, for the college's staff and facilities were unprepared for both the number and the needs of these new students. As a result, Shaughnessy's efforts to shape a program of instruction took place under chaotic conditions, with many classrooms merely partitioned areas of the college's Great Hall and with last-minute hirings of teachers when and if budget lines became available. The measured tone of Shaughnessy's account of the impact of Open Admissions in her writings distances the pressures and responsibilities that pushed her close to exhaustion and led to a brief physical collapse in 1971. Her frequent allusions to the pioneer role of basic writing teachers and to the "frontier" experience of such work had more to do with her sense of taxing work loads than with nostalgia for her Western past.

Nevertheless, the frontier image had another application, one Shaughnessy referred to explicitly when she used it. The frontier involved uncharted territory, and probably the absence of any familiar markers or guideposts was foremost in her own mind during the first years at City College. Few members of the faculty had worked with such students, there were no resources of experience to turn to elsewhere in the university, and there was neither opportunity to contemplate the problem nor time to plan to meet it. The frontier experience, with its mixture of pragmatism and idealism, defined Shaughnessy's situation; she educated herself by trusting in and observing her students, by raising with her colleagues simple questions about what students needed to learn, questions that had no simple answers.

Because Shaughnessy's approach was largely inductive, the City College writing program was not shaped by a governing rhetorical or linguistic theory but was developed through observation and research seeking to make sense of those observations. The most telling example is Shaughnessy's own study of thousands of proficiency essays written by entering students, a project that eventually resulted in *Errors and Expectations*. In that book, her observation that writing teachers find "a repertoire of approaches" useful and usually evolve "a Rube Goldberg grammar, full of borrowed and makeshift parts, unsupported by an overarching theory" illuminates her own development as a teacher and scholar.[5] Instead of establishing a required curriculum for the writing program, she encouraged teachers to follow their hunches and share their insights with one another, and she encouraged them as well to engage in a wide range of research projects: studies of derailments in student prose, contrastive studies of first language interference among nonnative speakers, and examinations of perceptual problems that affect some students' ability to proofread.[6] She also sponsored a different kind of project that sent English teachers as auditors into introductory courses in disciplines unfamiliar to them, such as biology and psychology.[7] Their efforts to grasp the concepts governing these subjects made them more aware of the particular intellectual assumptions and the distinctive languages appropriate to these disciplines. Transforming teachers into learners, a constant in Shaughnessy's pedagogy, but here done quite literally, made the teachers comprehend the situation of students new to all kinds of academic discourse.

As these projects began to yield more information about underprepared students' behaviors as learners and writers, Shaughnessy also developed programs at City College for teachers: first an in-service workshop for adjunct teachers, then a Master of Arts program in the teaching of writing, and finally a special seminar for teachers who, because of declining enrollments, were being reassigned from other disciplines to composition teaching and who needed retraining. By

1972, Shaughnessy was active in the founding of the CUNY Association of Writing Supervisors and was a speaker at national conferences on the teaching of writing; by 1975 she had begun to serve as a consultant on basic writing to governmental agencies, to foundations, and to associations of English teachers. In the fall of 1975, she left City College to become associate dean for academic affairs in the City University, as well as director of the CUNY Instructional Resource Center, a position that enabled her to continue to develop the research interests she had initiated in her college work. Although she served as dean for only two years before illness began to curtail her activities, in that time she supervised a detailed review of basic skills programs at all the CUNY colleges and laid the groundwork for the assessment tests in writing, reading, and mathematics introduced by the university in 1978. The first of these projects reflected Shaughnessy's belief in educational diversity and eclecticism ("there are many ways to climb Mt. Fuji," she once wrote);[8] the second, her insistence on educational accountability and the achievement of competence by basic skills students. At the center, she also founded *Resource,* a newsletter of current information for basic skills teachers, and she brought with her from City College the *Journal of Basic Writing,* a scholarly periodical she had helped to establish a year earlier. The *Journal* focused on the academic issues that most concerned Shaughnessy, and much of the work it publishes today continues to reflect her intellectual influence.

These accomplishments are summarized here because they show how Shaughnessy, when she guided a writing program at City College for nearly 3,000 and was also involved with a tutoring center serving another 600 or 700 students, or when she later was responsible for basic skills in one of the largest university systems in the country, constantly made her administrative work into a scholarly exploration. Once again, her dual commitment enriched basic writing programs, for it encouraged teachers to combine a humane response to students' needs with a detached and disciplined study of the features of students' writing. Shaughnessy, in effect, made basic skills into a serious academic program; she sustained the morale of teachers during difficult times by virtually creating for them a professional field that required the same concern for hypothesis and proof that marked traditional academic inquiry. The City College writing faculty became aware that their program was a model looked to by many other colleges, as English Departments began to learn from Shaughnessy, directly or indirectly, who basic writers were and how they should be taught.

At City College, then, Shaughnessy gained experience in two roles that had an important influence on her writing. In one, she educated and encouraged younger teachers and convinced many of them that basic writing could serve as a challenging academic career; in the other, she persuaded older teachers of literature that basic writing students could be taught if teachers were willing to readjust, not their standards, but their methods of teaching. "Diving In," Shaughnessy's most notable address, exemplifies her way of seeking to mediate between teachers, young and old, and their new constituency. As one might expect, she proceeded analytically, classifying and ordering, but now applying a schema for developmental learning to teachers rather than students. Her talk described four stages of development for a teacher encountering a basic writing class for the first time. The first stage, "Guarding the Tower," is characterized by despair at the alarming deficiencies of basic writers; at the next stage, "Converting the Natives," the teacher sees these deficiencies as simple ignorance and presents systematic instruction in grammar that the students fail to absorb. In stage three, "Sounding the Depths," the teacher begins to recognize the intelligence of these students and to realize that their problems as writers can only be understood by observing their writing carefully instead of insisting on preestablished forms of instruction. The fourth stage, "Diving In," is less a stage than, as its title implies, a gesture of commitment. It asks for belief in the basic writing students' incipient excellence, it reminds teachers of their role in a democratic society, and it offers the academic challenge of a new field of research. These stages of development almost certainly chart Shaughnessy's own experience in the late 1960s of confronting a totally new teaching situation, being forced to abandon familiar teaching methods, and retraining herself through careful observations that led to new hypotheses.

"Diving In" never draws lines between opposing camps, never issues challenges or denounces incompetence. Instead, Shaughnessy uses her method of classification to introduce the idea of gradual change, step-by-step adjustments that lead the teacher from one perspective to another. That process, moving from preconceptions based on a familiar world to the wrenching effort to change one's usual responses and meet a new situation, applies to both student and teacher as Shaughnessy makes clear when she offers a developmental education to the "disadvantaged teacher."

Although "Diving In" typified Shaughnessy's efforts to avoid contention by directing people toward responsible action, she did not sacrifice her convictions in order to conciliate those who continue to "guard the tower." In fact, a passage from one of her annual reports to the English Department offers an example of outspokenness that few administrators would choose (or would be able) to duplicate:

> We begin to realize how difficult it is not to become a high school of the kind many of our students come from as so many of the conditions that plague them begin to plague us—institutional cynicism which leads admin-

istrators to plan for numbers rather than people and cultivate myths about how people, or at least certain people, don't have to be able to write any more or how some other short-cut scheme like mass lectures or automated instruction will solve problems that are rooted deep in the miseducation of our students and the sins of our society; in the attitude of teachers who lose the sense of their students and find (and rationalize) their own short-cuts, perceiving themselves as being violated rather than their students as being betrayed.[9]

Shaughnessy's anger at academic complacency or injustice is evident in such an indictment; and one of her final talks, "The English Professor's Malady," looks critically and somewhat bleakly at the elitism of most English Departments and the indifference of their faculty toward the basic writer. But her gift was always for seeing the potential in people—teachers as well as students—and for facing new tasks rather than condemning past failures. "Diving In" is characteristic of her willingness to build bridges rather than barricades and to sketch out the process by which English teachers can come to devote their best energies to literacy as well as literature.

Shaughnessy's only book, *Errors and Expectations*, embodies her effort to go beyond the encouragement she offered teachers in "Diving In" and to suggest specific ways in which, having made this commitment to basic writing, they can keep themselves afloat. A reader encountering *Errors and Expectations* for the first time will be struck by Shaughnessy's distinctive way of introducing basic writing students. She provides no elaborate explanation for their academic failings, nor does she do more than allude to the inequities of the school system in which they have struggled to survive. Instead, her portrait of these students returns consistently to two assumptions. The first is that they are highly motivated, a point she supports by turning paradoxically to their record of academic failure. Only students who value learning, she says, would return again to the classroom, the site of twelve years of frustration and embarrassment. Once we accept the idea of the student who wants to learn, the burden of failure falls on the schools: The errors teachers see in students' writing "raise the question, again and again, of what these students got for the twelve years they gave to schooling." The record of failure also reveals the students' second major strength. For Shaughnessy, most of their writing errors indicate an effort to treat language logically; errors are most often "the result not of carelessness or irrationality but of *thinking*" (p. 105) and reflect the students' distrust of haphazard guesswork and their desire for predictability. These two assumptions—that basic writing students want to learn to write and that they can supplement the intuitions they possess as native speakers with rational powers of mind—are also the basis of Shaughnessy's response to the anxieties of teachers. For before she turns to any matters of pedagogy, she defines the basic writer as motivated and rational—in other words, as eminently teachable.

Another point that will strike her reader, for it is the most dramatic and shocking element of the book, is its documentation of the kind of writing that basic writers actually produce. The unfamiliarity of such writing to all but basic writing teachers creates this shock as it did when the book first appeared. For the writing of these students, with all its inadequacies, was rarely represented in public print except for the occasional sample quoted by a despairing faculty member interested in proving such students ineducable or by an angry legislator interested in proving the professor incompetent. The writing itself is shocking because the errors are so fundamental while, at the same time, the young adult writers are so obviously more advanced in the complexity of their thought than grade school children writing for the first time. *Errors and Expectations* surrounds us with examples of basic writing:

> But many colleges have night classes so you could have worked and gone to college also pay for your education although some other programs to help pay on some where you don't pay or some where you don't pay at all so you were lazy [p. 53].

> Education is a mean to the end although some people have their education and play dum that doesn't mean that we the younger generation should just set and watch other people we want and education we got to fight for it, it is not easy but one person told me that anything that is easy to get is not worth while having [p. 20].

> I feel that For a young person. Who has Just completed High School and Wishes to attended College. To get a higher Education that this moved is a very wise one. I know For a Fact that there are people. Who have attended college and have received a college degree. Who are reciving the same paid as a high school graunted. To me I am attending college because I think it is a beutiful expreince. I also feel that it is a chance of a lifetime. To get more out of life and to better things for you in the Future [p. 28].

Even textbooks addressed to the needs of basic writers gave one no sense of the sheer tangle of forms produced by the inexperienced adult writer; textbook samples of error always represented a single kind of breakdown within the sentence, correctable by transforming one element through the application of one rule. Shaughnessy gives her readers a profusion of error, example after example, in passages that often require repeated reading before more than a glimmer of the writer's meaning can be seen. She can expose such writing to her readers because of her conviction that all errors, except those caused by simple carelessness, can be interpreted within a framework of purposive, even if mistaken, applications of language. The sheer number of such examples in her book is necessary to demonstrate that they are generally amenable to pattern and rule, that she is not making highly selective choices from a mass of incoherent and unsalvageable material.

The disorder of student error is balanced by the way Shaughnessy responds to such errors. Here is a single student sentence and her subsequent comment:

> The most disadvantage and disappointment is knowing and hoping that somehow the field which one chose does not have an opening after college.

> Here the writer intends to speak of two feelings—a feeling of disappointment at discovering that even with a college education there is no job and a feeling of hope that there will be a job. The gerund *knowing* needs to be completed by one clause (that the field does not have an opening); the gerund *hoping* by another (that there will be a job). Thus the consolidation the writer attempts by yoking *knowing* and *hoping* and then attaching them to the same complement cancels out his meaning [p. 52].

Shaughnessy's serious attention to the sentence she quotes transforms it from failed prose to potentially successful prose and immediately divests it of its power to shock. She does not ignore its deficiencies, but remains unruffled by them. Instead, she responds carefully and directly to the writer's intention and explains why the sentence has not realized that intention. The tone and method of her analysis (undoubtedly shaped by her professional experience as a copy editor) would be equally suited to a piece of publishable prose, although her purposes are obviously different. Though Shaughnessy never glosses over the inadequacies of her examples, she brings the basic writing students into the academy by regarding what they say as the work of genuine, if untrained, writers.

Shaughnessy's passion for order and clarity is everywhere evident in the book. Errors are classified, common rules of grammar are arranged in manageable patterns, and significant types of instruction are listed in sequences that might best meet student needs. The chapter divisions and the orderliness of the presentation recall the procedures of conventional handbooks, offering a systematic review of grammatical rules, but here the orderings sort out the overlapping functions of words or syntactical arrangements that often confuse the basic writer. For example, Shaughnessy devotes several pages to a discussion of the word *it*, prefaced by the remark that "teachers tend to lump all *it* errors into one group, even when the causes and solutions of those errors may be very different." The elaborate sorting process that follows concentrates not on the familiar pronominal slots, but on how a writer makes use of *it* "as a free-floating substitute" or as a way "not simply to re-emphasize or classify his subject but to recapture it" (pp. 69–70).

The grammatical issues that Shaughnessy chooses to discuss are often those that frustrate teachers because of their apparent simplicity and pervasiveness. The *-s* inflection that marks nouns and verbs illustrates the kind of grammatical feature that, when not controlled by the writer, produces a high incidence of error because the form recurs so often; moreover, the error can attach itself to simple lexical items (*he say, two boy*), creating the impression that the writer lacks rudimentary skills. By now, most teachers are aware of how dialect and phonology explain the student writer's uncertain use of this inflection, but Shaughnessy pursues the matter much further:

> At least two other reasons for the -s difficulty must be added to the phono-logical: that the -s serves no purpose, the number of the subject already being indicated by the subject itself or by a limiting adjective; and that the stem form of the verb often appears after a third-person subject in con-texts that are difficult for a learner to distinguish from the third-person present indicative. The first reason suggests why students have so little motivation to master the form, and the second reason suggests why they have so much difficulty mastering it even when they try [pp. 95–96].

Such a passage exemplifies Shaughnessy's method: It contemplates grammatical pattern from the perspective of its multiple misuses in hundreds of student papers and thereby recognizes not the rule's au-thority, but its susceptibility to misconstruction. As the passage sug-gests, these misconstructions can be understood by considering such variables as the way familiar speech and hearing patterns affect written performance, the way a rule may run counter to a student's sense of consistency, or the way specific applications of the rule can produce forms easily mistaken for the applications of another rule. Thus, the simple rules of pronoun form or verb inflection are transformed by seeing them from the perspective of untrained writers.

As Shaughnessy adopts this new perspective, she modifies conven-tional patterns of classification in two important ways. First, her discus-sions are arranged according to a hierarchy of error; within its chapters, her book pursues problems the teacher is likely to encounter frequently and sorts them out according to what the student writer may be attempting to do. The chapter "Syntax," for example, classifies one sequence as blurred patterns (the student combines features from sev-eral syntactic patterns), consolidation errors (the student attempts to express complex ideas by adding to the base sentence), and inversions (the student attempts to rearrange the customary pattern of the sen-tence). Secondly, classification no longer involves a series of rules and exceptions, but instead the identification of the variables that can lead the student writer to a particular form of incorrect expression. Shaugh-nessy at one point describes grammar as "a web, not a list, of explanations" (p. 130) and elsewhere discusses "the web of discourse"; these images of interwoven elements govern her descriptions and make classification such a flexible instrument for her purposes.

Shaughnessy, therefore, asks the teacher not to think of error as random failure, but instead to accord it a logic of its own. (*The Logic of Error* was her original title for the book.) Her assumption is that error, because it is based on language experience and rational needs and not on sheer ignorance or perversity, is deep-rooted and will change only slowly and through systematic instruction. Thus, she marks off three separate goals of instruction—awareness, improvement, and mastery— in order to establish different moments of satisfaction for both teacher and student. She also creates patterns that reflect the need for extended experience with a particular form so that the student can first grasp the nature of the problem—"Grammar should be a matter not of memorizing rules or definitions but of thinking through problems as they arise" (p. 137)—and then, through a range of sentence exercises, recreate and generate correct forms, ultimately achieving mastery in their use. Lengthy and involved as it is, such instruction acknowledges, not with despair but with determination, the kind of resistance it faces and sees, in the occasions for contemplation and reconsideration that writing offers, a chance for students to engage in a more careful and informed scrutiny of their own work.

At the same time, it must be acknowledged that the instructional models Shaughnessy provides represent a familiar and even conservative pedagogy. Quite rightly, she saw as the most pressing issue the need to understand the new students and to fathom the complexities of their errors. Her suggestions about how to respond systematically to these errors, though explained with great lucidity, never attempt to redefine instruction in the same way her work redefines the meanings of unskilled writing. Shaughnessy undoubtedly believed that the transformation of a teacher's attitude would in itself revitalize a teacher's methods; as a result, the specific instructional sequences described in *Errors and Expectations* remain less compelling than the book's broader discussions of a teacher's tasks and goals.

Whatever reservations one may have about particular grammatical lessons, Shaughnessy's book strengthened the position—one that was being debated at the time the book was published—that writing instruction must attend to the matter of error. Many supporters of open access to higher education were arguing that a concern with error was the mark of discredited prescriptive modes of instruction, modes that discouraged students and deprived them of pleasure in writing and trust in their own creativity. Shaughnessy made a new case for grammar instruction because the student writing she reproduced made self-evident the need for some intervention to reduce the incidence of error so high that it repeatedly cripples the writer's ability to communicate. What seemed to radical teachers of the 1970s a reactionary attitude was, in fact, one that sympathized with the aspirations of students, aspirations that would be blocked by their inability to write. The surmounting of error, therefore, stands as one part of Shaughnessy's larger concern with basic writing students' need to gain control of a language that would enable them to participate more fully in a wider society. She would not contest "the students' right to their own language" (the theme of a resolution adopted by CCCC in 1974), but she would insist that more than one language was necessary, and the second language should be the public discourse that the schools should assist us all to master. Her work reflects the values expressed by Richard Hoggart when he protests that some fellow radicals are not sensitive to the real needs of their students:

> We cannot leave people in corners, having to our own satisfaction redefined those corners as nicer than the outside, more public world. We are talking about something different from training people to acquire bourgeois speech and assumptions.... We are talking about having that respect for them which requires us to help them gain greater, more articulate, and more self-conscious access to their own personal and social lives.[10]

Shaughnessy demonstrates the kind of help to which Hoggart refers, for her book enacts the virtues and powers she ascribes to academic discourse. In prose that is itself scrupulous in its attention to evidence, attuned to a skeptical but reasonable audience, and committed to what she calls "high standards of verification and sound reason," she defends the basic writer. Her attention to matters of tact, courtesy, and formality give her writing a dignity and seriousness appropriate to the importance of her subject. In her manner, as well as in the substance of her book, Shaughnessy differs from many other writers who at the time took issue with the traditional methods of higher education. Their attitudes were customarily announced by marked changes in tone and language from earlier books on the teaching of writing. Peter Elbow, whose influential work *Writing Without Teachers* appeared in 1973, provides a characteristic example of "anti-academic" discourse as he advises his readers thus:

> If you want to write, you must cook. There is always a crunch in writing. The crunch feels to me like lifting the Empire State Building, like folding up a ten foot parachute on a windy field. You can't avoid the crunch. It takes heat, electricity, acid to cook. If you can't stand the heat, get out of the kitchen.[11]

The highly personal note that Elbow strikes, the casual manner, the use of a decidedly untraditional vocabulary to describe writing experience ("cook" "crunch") all serve to announce his disaffection from the academic world. Elbow's purposes are, of course, different from Shaughnessy's—he addresses the writer, not the teacher—but his prose shows by contrast how measured and controlled Shaughnessy's writing is. Her decorum, like his rejection of it, constitutes a political statement.

Although Shaughnessy demonstrates her faith in the academy in her own language and method, that faith is actually based on a belief in the academy's ability to renew itself. She insists that the institution is not sacrosanct and may not complacently restrict itself to students who measure up to its standards. In a profession in which virtually every teacher of note has published a textbook (although, Shaughnessy once remarked, they rarely represent "the best energies and motives of their authors"[12]), she never did;[13] her writings, though they were grounded in her experience with students, were always addressed to teachers and administrators. She chose this audience because of her conviction that educators were the ones to be educated. As she repeated in virtually every talk and article, the new students would learn if teachers would change and if higher education could free itself of "the illusion that the new society will be bounded by the values and interests of the white, privileged, male minority that has called itself Western Civilization."[14] Her moral convictions caused her to question self-serving definitions of the humanities, usually offered by faculty who wished to remain undisturbed, saying familiar things to familiar students. Here is one of her comments on a City College proposal defining the nature of the humanities:

> The new students, in order to mobilize their energies need to see connections between their courses and their careers. This troubles many humanists who fear a debasement of the educational experience in the pursuit of a career curriculum. "The humanistic studies of the university," says the proposal, "are the means by which the student will ultimately be able to evaluate himself as a human being rather than as a professional or vocational performer." But is it not in our professional and vocational roles that we need to evaluate ourselves as human beings? Is it not in our work that we discover the real limits of our humanity? Is it not our inability to extend generosity and imagination and responsibility beyond our private lives that creates so much of the cruelty and shoddiness of our public lives? It is obvious from all the data we have, not only from minority students but from open admission students generally, that professionalism and paraprofessionalism are the routes the new students will be seeking through academia. The redemption of the world of work in this society may depend upon a redefinition, a broadening, of the concept of education to include more than the narrow intellectualism that has often resulted from what we call a humanistic education.[15]

Shaughnessy's words here demonstrate another distinction between her perspective and that of other critics of traditional pedagogy. Many dissident academics in the 1970s expressed their disillusionment with American society and its institutions by redefining training in writing as predominantly the creation of a personal voice, one that resists the anonymity and dehumanization they associated with public address. But Shaughnessy never granted the private self such a privileged position, independent of society, even though her involvement with basic writing students made her fully aware of the inequities of American life and the human cost of those inequities. To control the codes of public discourse is to participate in society and to have the choice of how or on what terms one wants to participate. When Shaughnessy painstakingly works out the various explanations of the -s inflection, she is not resolving a localized problem in grammar, but contributing to the redemption of "what we call a humanistic education."

Despite the breadth of her statements on social change and higher education, Shaughnessy never argued for dramatic changes in what should be taught to student writers; her concerns covered the most familiar categories in writing pedagogy: punctuation, syntax, vocabulary, spelling, and the customary forms of essay organization. More surprisingly, she never developed a theory of composition to direct the novice teacher. She distrusted theoretical completeness or what she called "The Way or The Book or The Grammar," feeling that a teacher committed to a system would be correspondingly inattentive to the specific needs of individual students. Although her own methods of explanation rely heavily on classification, all her work resists closure; instead, it looks to the future, emphasizing what needs to be learned and done. The operative word in *Errors and Expectations* is "tasks," not "achievements," and the last essay published in her lifetime was titled "Some Needed Research on Writing."

Shaughnessy's own achievements, as we have seen, were both substantial and lasting. She demonstrated that basic writing students belong in college, and she provided a way for teachers to take up that argument with greater confidence and to address the needs of their students with greater effectiveness. She also demonstrated that basic writing teachers themselves belong in college, for she honored their work, defined the complexity of its challenges, and created models for educational research appropriate to a professional field. Most important, she stirred her profession by centering her attention on deeply flawed texts, drawing from them an ideal student and an ideal teacher, both committed to learning in its fullest and richest sense. Undoubtedly, most students and teachers fall short of that ideal, but in her book Shaughnessy never wavers in her belief. The extraordinary "readings" of the unreadable that constitute the single most vivid achievement of *Errors and Expectations* serve as her proof that her ideal student is already there, because her claim for the basic writing student is an ideal of readiness and potential, not of accomplishment. With that established, she can then challenge teachers to fulfill their complementary ideal role as defined by a truly humanistic tradition. Academic life in the early 1970s was certainly touched by idealism, but not always by the kind that valued observation and research and expressed itself with elegance and restraint. The moral passion that led Mina Shaughnessy to look so carefully at so many thousands of student essays, more than any data or insight or program that resulted, constitutes her most important legacy. That passion makes her work not only admired by writing teachers but also cherished.

Notes

I would like to thank Sarah D'Eloia Fortune, Marilyn Maiz, Edward Quinn, and Alice Trillin for their recollections of Mina Shaughnessy's time at City College, and Marilyn Maiz also for her help with the bibliography.

1. Maurice Hungiville, "Mistakes in Writing: Symptom or Sin?" *The Chronicle of Higher Education*, 4 April 1977, p. 18.
2. *The Nation*, 9 December 1978, p. 645.
3. Press Release, *National Endowment for the Humanities*, Summer 1978.
4. *John D. Rockefeller: A Portrait* (New York: Harper, 1956); *Adventure in Giving: The Story of the General Education Board* (New York: Harper & Row, 1962).
5. *Errors and Expectations* (New York: Oxford University Press, 1977), p. 156. Subsequent citations from this work will be included parenthetically in the text.
6. *Journal of Basic Writing*, 1:1 (Spring 1975), provides evidence of the range of work being done at City College because every article in that initial issue of the journal is by a teacher in Shaughnessy's program.
7. Unfortunately, the project had to be abandoned after a year because of lack of funds. The reports generated by the project have never been published.
8. *Journal of Basic Writing*, 1:2 (Fall–Winter 1976), p. 1.
9. Chairman's Report, 1973–74, Department of English, City College of New York, p. 4.
10. Richard Hoggart, "The Importance of Literacy," *Journal of Basic Writing*, 3:1 (Fall–Winter 1980), p. 85.
11. Peter Elbow, *Writing Without Teachers* (New York: Oxford University Press, 1973), pp. 66–67.
12. "The English Professor's Malady," *Journal of Basic Writing*, 3:1 (Fall–Winter 1980), p. 92.
13. In 1975, Shaughnessy made a contract with Scott, Foresman and Company to produce an English handbook, but she withdrew from the contract a year later.
14. "Humanistic Studies and the New Students," unpublished manuscript.
15. Ibid.

Selected Publications of Mina Shaughnessy

Note: Asterisked entries were published under the name Mina Pendo.

*"Milton" (poem). *The Hofstra Review*, 2:1 (Spring 1967), p. 23.
*"Reason Under the Ailanthus." In *Washington Square by Henry James*. Ed. Gerald
   Willen. New York: Crowell, 1970, pp. 243–252.
   Sees Catherine Sloper as a victim, caught between her father's chilly
   objectivity and commitment to reason and her suitor's shallow self-inter-
   est. Catherine's lack of vitality diminishes the force and interest of the
   novel.
"Open Admissions and the Disadvantaged Teacher." *College Composition and
   Communication*, 24 (December 1973), pp. 401–404. Reprint in *Journal of
   Basic Writing*, 3:1 (Fall–Winter 1980), pp. 104–108.
   Argues that "the alphabet of numbers" (test scores, grade-point aver-
   ages, attrition rates) deflects teachers from looking carefully at individual students and the writing they produce. Sees how the presence of
   disadvantaged students will raise significant questions about the purposes and practices of higher education.
"Speaking and Double Speaking About Standards." In *Proceedings of the California State University and Colleges Conference on the Improvement of Writing
   Skills*, Spring 1976, pp. 79–83. (Initially a speech delivered at the conference, June 3, 1976.)
   Distinguishes between invoking standards to exclude all but the most
   competent students from college and exploring the real nature of standards in order to help less skilled students master what they need to
   know. Much of the material in this paper appears in Chapter 1 of *Errors
   and Expectations*.
"The Miserable Truth." In *The Congressional Record* (September 9, 1976). Reprint in *Journal of Basic Writing*, 3:1 (Fall–Winter 1980), pp. 109–114.
   (Initially a speech delivered at the First Annual Conference of the
   CUNY Association of Writing Supervisors, April 26, 1976.)
   Describes the erosion of the Open Admissions program at the City
   University of New York owing to financial constraints and administra-
   tive indifference. Celebrates how much basic writing teachers have
   learned and achieved and insists that, regardless of temporary setbacks,
   Open Admissions has permanently changed the character of higher
   education.
"Diving In: An Introduction to Basic Writing." *College Composition and Communi-
   cation*, 27 (October 1976), pp. 234–239. Reprint in *The Writing Teacher's
   Sourcebook*. Ed. Gary Tate and Edward P. J. Corbett. New York: Oxford
   University Press, 1981, pp. 62–68. (Initially a speech delivered at the
   Modern Language Association, December 1975.)
   Traces the development of a teacher's response to basic writing students from an initial disbelief that they are educable to a perception of
   their incipient talents and a commitment to discover new and more
   effective methods of teaching them.

"Basic Writing." In *Teaching Composition: 10 Bibliographical Essays.* Ed. Gary Tate. Fort Worth: Texas Christian University Press, 1976, pp. 137–167.

A bibliographical essay citing and commenting on published works of importance to the basic writing teacher. Section 1 describes publications that discuss basic writing students, the characteristics of their writing, and effective methods of instruction. Section 2 surveys publications in linguistics and in research on writing behavior that would be of particular interest to the basic writing teacher.

Errors and Expectations. New York: Oxford University Press, 1977.

A detailed study of the most common writing difficulties of basic writers, analyzing the causes of these difficulties and suggesting principles and practices to guide the basic writing teacher. The book takes up such categories as handwriting and punctuation, syntax, common errors (chiefly grammatical inflections), spelling, vocabulary, and significant patterns beyond the sentence. This is Shaughnessy's major work.

"Some Needed Research on Writing." *College Composition and Communication,* 28 (December 1977), pp. 317–320. Reprint in *Journal of Basic Writing,* 3:1 (Fall–Winter 1980), pp. 98–103.

Asks for more precise observation and analysis of growth in writing skills, characteristics of academic writing, specific subskills writers need to learn, and teaching effectiveness in the classroom.

"Statement on Criteria for Writing Proficiency." *Journal of Basic Writing,* 3:1 (Fall–Winter 1980), pp. 115–119. (Prepared in November 1976 as a working paper for the CUNY Task Force on Writing.)

Divides the assessment of writing into two areas of competence: "givens" or correct forms and "choices" or the selection of words, sentence patterns, and rhetorical structures.

"The English Professor's Malady." *Journal of Basic Writing,* 3:1 (Fall–Winter 1980), pp. 91–97. (Initially a speech delivered at the Association of Departments of English Conference, Albany, N.Y., June 1977.)

Criticizes English Departments for their lack of interest in writing instruction and lack of knowledge about the literature concerning this subject. Warns of the dangers of an elitist attitude in a democratic system.

Halpern, Jeanne W., and Dale Mathews. "Helping Inexperienced Writers: An Informal Discussion with Mina Shaughnessy." *English Journal,* 69 (1980), pp. 32–37.

Comments on the ways in which teachers need to change their attitudes and revise their expectations concerning basic writers.

CONTRIBUTORS

Ann E. Berthoff, professor of English at the University of Massachusetts, Boston, has written on the theory of metaphor and Renaissance poetics in *Sewanee Review, Review of English Studies, Modern Language Quarterly,* and other journals, and is the author of *The Resolved Soul: A Study of Marvell's Major Poems.* She has published three books for writers and teachers: *Forming/Thinking/Writing; The Making of Meaning;* and *Reclaiming the Imagination: Philosophical Perspectives for Writers and Teachers of Writing.*

John Brereton is associate professor of English at Wayne State University, where he directs the writing program. He has taught at Rutgers University, the City University of New York, and Columbia University. Professor Brereton has published articles on composition instruction and is the author of a rhetoric text, *A Plan for Writing,* and a forthcoming collection of readings for composition.

Wallace Douglas is professor emeritus of English and education at Northwestern University, where he taught since 1945. He has written on Wordsworth and on literary criticism, but his chief publications have been in the history and theory of composition. Professor Douglas directed the Northwestern Curriculum Study Center in 1968 and served as chair of the Conference on College Composition and Communication in 1969.

Walker Gibson, professor of English at the University of Massachusetts, Amherst, has taught at Amherst College and New York University. He

190

has published articles, poems, and reviews; his books include *The Reckless Spenders, Seeing and Writing, Tough, Sweet, and Stuffy,* and *Persona.* In 1971–72 Professor Gibson served as President of the National Council of Teachers of English.

William Irmscher is professor of English at the University of Washington, where he directed the freshman writing program for twenty-three years. He served as editor of *College Composition and Communication* from 1965–74 and chair of the Conference on College Composition and Communication in 1979. His book, *Teaching Expository Writing,* is a standard reference for teachers of composition. Professor Irmscher teaches graduate courses in current rhetorical theory and theory of composition.

Richard Lloyd-Jones is professor of English and director of the School of Letters at the University of Iowa, where he has taught since 1951. Professor Lloyd-Jones was a co-creator of the Primary Trait Scoring method and a co-author, along with Richard Braddock and Lowell Schoer, of *Research in Written Composition.* He has served as Chair of the Conference on College Composition and Communication and is currently president-elect of the National Council of Teachers of English.

Robert Lyons is associate professor of English at Queens College, City University of New York, where he has served as director of the composition program. During 1979 Professor Lyons was director of the Instructional Resource Center at CUNY. His publications include *Language and the Newsstand* (with Thomas Van Laan) and *Autobiography: A Reader for Writers.*

Donald Stewart, professor of English at Kansas State University, was chair of the Conference on College Composition and Communication in 1983. He has published many articles on composition and rhetoric and has contributed chapters to *The Present State of Scholarship in Historical and Contemporary Rhetoric,* and *The Rhetorical Tradition in America.* He is currently at work on a book-length study of Fred Newton Scott's life and professional career.